How to Build a Puppy

HOW TO MANAGE A MESS

How to Build a Puppy

...into a healthy adult dog

Julia Robertson

Galen Myotherapy

CRC Press
Taylor & Francis Group
Boca Raton London New York

CRC Press is an imprint of the
Taylor & Francis Group, an **informa** business

A TAYLOR & FRANCIS BOOK

First edition published 2022
by CRC Press
6000 Broken Sound Parkway NW, Suite 300, Boca Raton, FL 33487-2742

and by CRC Press
4 Park Square, Milton Park, Abingdon, Oxon, OX14 4RN

CRC Press is an imprint of Taylor & Francis Group, LLC

ISBN: 978-1-03-221521-1 (hbk)
ISBN: 978-1-03-221520-4 (pbk)
ISBN: 978-1-003-26878-9 (ebk)

DOI: 10.1201/9781003268789

Typeset in Palatino
by Deanta Global Publishing Services, Chennai, India

To my father, Albert Clifford, who has always inspired me

Contents

Preface

A LIFETIME OF LEARNING AND OBSERVING

All my life I have worked with animals, observing their physicality and behaviour. For more than 20 years I have been successfully observing and treating dogs suffering hugely from ongoing muscular problems. The thousands of dogs I have treated over the years have presented with symptoms and behaviour that very often have been difficult, have not been diagnosed with standard diagnostics, or did not respond well to current medication (Figure 0.1). Many of the dogs I saw had secondary conditions such as osteoarthritis and other skeletal issues, and often various behavioural problems too.

In the beginning I could not work out why so many dogs were displaying so much pain and discomfort, so I began to study the correlation between what I was seeing and treating with exercise, activity, and environment. The connections between these three elements were alarming but it was surprising how very simple changes could be made to all of these and substantially improve the dogs' health.

SMALL CHANGES TO MAKE HUGE IMPROVEMENTS

It was all those years ago that I discovered we need to make changes to how we live with our dogs. We really needed to consider the dog's environment or the environment that we expected our dogs to healthily live within. Exercise and activities that we felt were making them happy and keeping them fit were often having devastating effects on their health.

The changes we need to make are relatively tiny and yet they will be profound.

WHY WAS THIS BOOK WRITTEN?

So many of our dogs are suffering, and because of how dogs express themselves, they are basically suffering in silence. Dogs do not appear to demonstrate discomfort or pain in a way that can be easily translated by us humans, so often our dogs are physically struggling, and we can inadvertently misconstrue what is actually happening within their bodies. Often, instead of demonstrating physical discomforts, they will have behavioural changes that frequently cannot easily be attributed to a physical issue, especially if it is chronic, or ongoing.

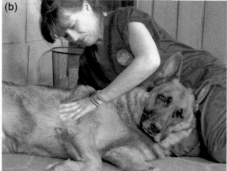

Figure 0.1 Julia performing hands-on Galen Myotherapy (see organisations, page 243) on two dogs showing signs of mobility deficits.

PAIN AND BEHAVIOUR

During this time of studying and learning from these dogs, it also became evident to me that there was a direct link between 'pain and behaviour'. This led to my understanding that making these changes would not just make a positive change to our dogs' physical health but also potentially to their emotional health (Craig, 2003).

The sad thing about all those dogs I had treated was that they were all damaged, and damaged muscles do not heal perfectly and will literally always carry the scars, so they all had to be managed for the rest of their lives.

PREVENTION IS BETTER THAN CURE

I was of course delighted to be able to enhance these dogs' lives, but prevention is better than cure, or management. Therefore, by using what I had learned from those thousands of dogs, this book was developed so we can start at the beginning and understand what we need to do to build a healthy puppy into a healthy mature dog.

THE PURPOSE OF THIS BOOK

We need to make small but profoundly important changes to the way we look after our puppies and our growing and developing dogs. Our environment and exercise and activity habits have deep effects on a growing puppy but also ongoing effects on our developing and mature dogs.

This book is primarily intended for new and developing puppies, but the fundamental advice is suitable and totally appropriate for all dogs, especially dogs that are rehomed or rescued.

It is also intended to demonstrate not only how we can make such small changes that will make an almost exponentially positive difference to our dog's health and life span, but also, and so importantly, how these changes and additions can be integrated easily into our normal lives, and how by understanding more about what type of exercise a growing puppy and dog needs, we can promote a more relaxed cohabitation between human and dog.

The book explains which movements and activities are beneficial to the development of your puppy and, also, as importantly, those that are detrimental and likely to create health problems throughout life.

Appropriate schedules that will ensure your puppy develops the three key components essential to a balanced structure are in Chapter 5, called **The Galen Puppy Physical Development Programme©**.

HOW WILL THIS BOOK HELP?

When we get a new puppy, they obviously appear perfect, and of course they are! However, sadly, just like many people, some puppies are born with a structure that is not perfect. Even if you get one from the best breeder, just sometimes, a puppy has not got the perfect conformation that was expected. Not only that, but birth can be quite a brutal event, and sometimes puppies can incur injuries from birth, or in the whelping pen or as they play vigorously with their siblings. All these things can be completely unintentional, but the effects can be already present in your new puppy.

This book will help you develop your puppy, in an all-inclusive way, even if they have not been blessed with the best conformation or the best start in life, or indeed suffered an unintentional injury or accident. Good examples could be the puppy that scrambled to get out of someone's hold when they were being carried and then was dropped to the ground. They appear to be unscathed, but they are not. This type of accident could have potentially created a soft tissue injury of some dimension that would develop negatively, exponentially, causing adaptive changes over their lifetime.

Applying these activities and observations could even help you identify issues earlier so that you can be aware and can make appropriate adjustments or have conservative treatment sooner.

Understanding how we can make some very simple changes to our homes, and how we exercise and train our puppies, will have a massively positive impact on our dogs' lives, which of course will have a positive effect on our lives, and most importantly, our lives with our amazing dogs.

THE STRUCTURE OF THIS BOOK

Personally, I do not like books that just give information without reason, fact or rationale, so this book has a dedicated large section to anatomy, pertinent to the physical movement of the puppy or dog.

The anatomy that is covered explains how everything fits together to form a functional moving structure, and how everything is interdependent.

This fundamental knowledge is, I believe, vital to understanding how the body works. Then the activity and home advice becomes more logical and easier to interpret, and therefore implement, using foundation knowledge, not just a tick box.

By explaining basic functional anatomy, it also becomes clear how the body parts cannot be considered in isolation. The dog's body operates as one static but also kinetic structure. All the working parts rely on the other working parts to form a healthy structure; it is not just the skeleton that holds us together, but the sum of all the working parts that are involved with locomotion that maintains our body's integrity. When our body's locomotion systems are working together, the body works in true synergy.

However, if the balance is not right, then the whole body will not function at its optimum and will start to adapt and alter its posture and loading. Slowly and insidiously, compensating and adapting until its 'weakest link' finally gives way.

This book was structured to make sense about why we need to help our puppies. It explains how a dog is built and how those working parts fit together. Then, once that is clearer, how we can get those working parts activated and functioning for our puppy.

It is then followed by suggestions of exercises and activities for your puppy and a programme that you can follow to help them build good foundations. Like all good structures, it is all about the good foundations.

It is hoped that having this understanding of how your puppy is put together, the programme will make more sense and give enough information so that you will feel confident to make adaptations and integrate these activities into your normal life. Most importantly, though, is to understand why some of the exercise routines and activities that have become almost 'traditional' are in fact incredibly damaging for our dogs.

KNOWLEDGE IS POWER

I believe that most of the people I meet genuinely want the very best for their dogs, but sadly, so many are misinformed or just did not know about their dogs' physical and emotional needs.

Ignorant seems a harsh word, but ignorance can also be not knowing there are questions to ask! I do feel that so often there is still so much 'old' information being circulated and that there is a genuine lack of good information to enhance dogs' physical and emotional health.

I hope this book will give you a greater understanding of your puppy and growing dog so you feel able to make good decisions based on knowledge to benefit your puppy, and also provide enough rudimentary knowledge to enable you to ask more questions from canine professionals and gain a much deeper understanding, which is empowering.

I want you to become aware of what your puppy needs for its body to develop into a strong, healthy, and robust mature dog, whilst giving clear information so that by making some very small changes, your home and your walks and activities will be productive and not destructive for your dog's body.

Acknowledgements

It is said it takes a village to raise a child, well it takes a really special group of people to write a book, and I am especially lucky to be part of such a group. Without their help, there is every chance this book would still be in unfinished chapters on my computer!

My family, as always, have been amazingly supportive. My children have consistently given their own wonderful inimitable support, and without it, I doubt I would have even had the belief in myself to start writing.

My incredible father, who painstakingly proofread every word I wrote, gave so much valued constructive help and suggestions. There were many occasions when I produced 'all the right words, but not in the right order', and he managed to untangle that maze of copy and produce exactly what I was trying to say. He also kept me motivated and on course, especially during the multiple times of self-doubt. I really do not know what I would have done without his help.

Thank you also to my friends and colleagues who are so dedicated to helping dogs, and also helping people to help their dogs, and whose passion for changing dogs' lives also kept me going.

Rosemary Sadler, who took so many incredible photographs for this book, which consequently helped make the copy descriptive and so much more engaging. Also, for the hours she spent going through endless photos to find that one special image!

Cushla Lamen, what can I say but everyone needs a Cushla in their life, a true and genuine friend, a great work colleague, and someone I highly respect. Thank you for your generosity of knowledge, support, and sharing with me the wealth of your experience.

Verity Halliday, thank you for your patience, plus your support in keeping the wheels turning with Galen whilst I was ensconced in writing, and your forbearance when being constantly told, 'I think I will be finished in a day or two…'!

Lucy Tyrrell for her gorgeous pictures of Marge, these photos really add to demonstrate the versatility of the programme. Allyson Giel for her wonderful academic support, and suggestions, which helped add more relevance to current thinking and behaviours.

Alice Oven (editor), your quiet, highly professional approach is one that is so easy to work with, offering just the right level of motivation, encouragement, and feedback that enabled me to get this book past the finishing line!

Turid Rugaas, thank you for asking me to write this book. For having the confidence and faith to entrust me to write about something you feel about so passionately. I am forever grateful that the stars aligned all those years ago, creating the crossing of our paths.

This book would never have been even started without all the dogs throughout my life that have inspired and taught me. I would like to thank the gorgeous Maggie and Tilly, who feature in the book, who always give without taking.

About the author

Julia Robertson established the Galen Therapy Centre in 2002 and since then has worked tirelessly to improve dogs' lives and their health. In this time, she has treated over 8,000 dogs and trained hundreds of people in Galen Myotherapy techniques. Julia was one of the very first in the UK to understand and treat the effects of adaptive change (or muscular compensation) in dogs, and through years of dedication has learned that trends, patterns of behaviour, and physical changes occur in a dog when they are suffering from muscular pain. This quantification of the nature of the changes has now been formalised into the Galen Comfort Scale©, which is being used in the many studies and treatments that Galen are involved in. Galen Myotherapy is also unique with its choice-led treatment methodology called Positive P.A.C.T.®, that Julia has developed through observations of dogs' behaviour during treatment, that allows them freedom of movement and choice through the treatment process. In late 2019, they changed their organisational name from 'Galen Therapy Centre' to 'Galen Myotherapy'. Julia speaks all around the world and runs an International Schools programme, and is co-author of another CRC Press book, *Physical Therapy and Massage of the Dog*.

Glossary

acute pain – generally sudden onset with a specific cause, by definition it does not persist.

anatomical plane– an invisible line used to describe the location of structures or the direction of movements.

aponeurosis – a flat sheet or ribbon-shaped connective tissue that anchors muscles to bones. It is similar to a tendon, but a tendon is more like a rope-type attachment as opposed to a more delicate and wider aponeurosis. It can also be considered a variant of deep fascia.

appendicular skeleton – the portion of the skeleton consisting of the bones that form the limbs.

atlas vertebrae – C1, the first vertebral body or bone of the neck.

autonomic nervous system – a component within the peripheral nervous system that helps our body to respond to environmental stimulation. It is divided into two main categories, sympathetic and parasympathetic nervous systems.

axial skeleton – is the portion of the skeleton consisting of the long axis, or skull, vertebral column (including the thoracic cage or ribs).

axis vertebrae – C2 the second vertebral body or bone of the neck.

brachycephalic – brachy = short cephalic = of the head. Short-headed dogs are prone to obstructive breathing due to the shape of their head.

central nervous system – the part of the nervous system that consists of the brain and the spinal cord.

cervical vertebrae – the vertebrae that form the neck.

choice – an evocative word, in this book context means giving the puppy the power of autonomy or agency, allowing freedom thought, which allows them to develop self-reliance and voluntariness.

chronic pain – ongoing pain, considered to usually last longer than six months.

clavicle/collar bone – a bone humans have to support the arms.

conditioning (muscle) – conditioning the body in this context is a programme that ensures the dog's body is physically connected through the locomotory systems to provide flexibility and strength for enhanced sustainability.

cross lateral movement – a movement that requires coordinating both sides of the body across the midline.

dewclaw – the digit on the dog's leg that on the forelimb relates anatomically to the human thumb. If they are present on the hindlimbs, they are often, but not always, without physical connection or function.

disc (vertebral) – positioned in between the vertebrae, acting as shock absorbers to the spine.

facet joints – the joints that connect the vertebral bodies that allow for movement, bending, and twisting, but also limiting movement.

fascia – a connective tissue that surrounds and connects the whole body. It is complex because its full function and form are still being discovered. It can appear visibly as thin as a spider's web or dense enough to stabilise the body. It is now found to be capable of sensations and actions, connecting organs to organs, as well as muscles to muscles and muscle groups to bones. It is also now thought to be a sensory organ recording where a body is in space (also called proprioception) as well as recording what is going on inside the body. Fascia was once thought to be tissue with no purpose, now more and more discoveries are being made, changing how we view the function of our body and how fascia is one whole-body connection.

> **It is generally acknowledged to be in three categories:**

- Superficial – located just below the skin.
- Deep – associated with connections of muscles and bones, along with some blood vessels and nerves.
- Visceral – mainly associated with connections of the internal organs.

feedback mechanism (physiological) – a complex physiological system of maintaining the body's equilibrium through stimuli and responses to those stimuli that correct imbalance. An example could be the regulation of body temperature, which dilates the blood vessels near the skin to allow heat to dissipate, or the regulation of oxygen, activating panting.

fight flight (sympathetic nervous system) – a layman's term for being in the sympathetic zone, or a heightened state of excitement.

foramen – an anatomical term meaning a hole in a structure providing a channel for other structures to pass.

foundation muscles – also referred to as deep muscles. A broad term for muscles that have little range of movement but provide deep stability for the joints primarily during locomotion.

functional anatomy – the relationship between anatomy and movement that anatomy can provide.

functional lines (movement) – relates to the movement lines the body uses to produce movement, through movement specific myofascial connections, therefore maximising muscle power through synergistic relationships of individual muscles.

gross anatomy – looking at anatomy as whole parts or structures, such as muscles or bones.

growth plates – the areas on a long bone where there is new bone growth. They are situated near the ends of long bones and are made of cartilage that goes through a process of ossification to form mature bone.

guardian – rather than owner. A name that is someone caring for and sharing their life with a sentient being. A dog being a sentient being should be part of a family group and not owned.

heat/inflammation – underlying inflammation can often be felt through the skin and can feel exceptionally warm to the touch. Inflammatory heat within this context could be from injury, overuse, or repetitive sprain

homeostasis – a definition of body balance or equilibrium from a physiological perspective. Greek meaning: homeo = same, stasis = steady, therefore a body maintaining a steady condition for optimal survival and health.

hyoid apparatus – the complex of bones that support the tongue and the larynx.

hyperextension – a joint being taken beyond the normal limits through additional force or forces.

innervate – supply the organ, part of the body with neural activation.

learned helplessness – a feeling of powerlessness arising from persistent stressful situations, failure, or trauma. Having negative effects on behaviour through the dog being unable to avoid or negate subsequent encounters (stressful, painful situations) with the same stimuli.

muscle compensation – injury or repetitive strain starts the cycle, then the postural and muscle changes create their own compensation, and the cycle continues, until appropriate treatment to adjust postural and muscle integrity (see Galen Myotherapy page 241, 243 and Figure 0.2).

muscle patterning – the coordination of muscle use within movement that maximises and engages the whole body using its functional anatomy to form good functional movement.

musculoskeletal – pertaining to the combined working of the muscular and skeletal systems.

myofascial – the combined working and interrelationship between the muscles and facial systems.

overload (myofascial) – when muscles are used beyond their functional use and can become damaged and/or change physiologically and structurally.

overuse (myofascial) – when muscle patterning is negatively interrupted, which creates an overuse within different muscles. It has many causes but often it is an imbalance between the foundation and power muscles.

owner – see guardian.

parasympathetic nervous system (rest and digest) – a specific division of the neurological system that controls the body at rest, facilitating digestion and helping relaxation.

physiology – the branch of biology that deals with the normal functions of living organisms and their parts.

posture – the position and alignment of the body when still or moving.

power muscles – also referred to as superficial muscles. A broad term for longer muscles that are generally positioned close to the surface of the body that have a large range of movement. Their primary role is to translate the dog's limbs into levers to provide locomotion.

proprioception – the perception and awareness of the movement and position of the body as a whole in space. For a dog, to have a good awareness of where all their feet are and being placed.

psychological – affecting or arising from the mind.

purpose (for a dog, in this context) – can be a breed fulfilment (e.g., gun dog fulfilling their breed purpose), and/or a role that a dog is happy to take on within the group that is fulfilling for them. This purpose fulfilment can often be formed with both guardian and dog, and can go towards a mutually fulfilling and productive relationship. As opposed to a purpose for a dog that is purely to make our (human) lives easier.

range of movement – the capacity of a joint to go through its complete anatomical range (this can be compromised through muscle dysfunction as well as joint dysfunction).

repetitive strain – repeated action or movement that has a negative impact on the soft tissue, creating an injury that can become chronic.

rest digest (parasympathetic nervous system) – a layman's term for being in the parasympathetic zone, or a relaxed or stress-free mental and physical state.

sentience – having the capacity to experience feelings and emotions.

spatial awareness – knowing where your body is in space in relation to other objects.

sympathetic nervous system – a specific division of the neurological system that rapidly controls the body's reaction and response to danger or stressful situations, including accelerating heart and respiration rates and redirecting blood to aid muscle function.

synovial joints – defined by having a joint capsule containing synovial fluid that is for joint lubrication. There are different types of synovial joints depending on the type of movement they allow.

systemic (body) – affecting the body generally.

vestigial – remaining in a form that is small or imperfectly developed and not able to function.

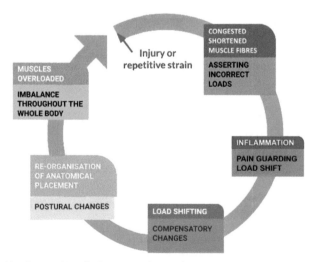

Figure 0.2 Galen Myotherapy Muscle Compensation Cycle©.

Introduction

ETHOLOGY AND FUNCTIONAL ANATOMY

Often, when we seek advice about how to look after a new puppy, we are told things to do and equipment to use, and often this is conflicting advice. If we take a step back and analyse what we are actually hearing, and ask ourselves, is the advice given primarily for the puppy's needs or ours?

Most advice appears to be designed to make 'puppy management' easier for us by controlling the puppy without consideration of the most important needs of your newborn family member.

When we are considering the best ways of looking after a new puppy, we should put ourselves into their 'paws' and consider our decisions from their perspective. There are so many direct comparisons with what is important to us, what is important to them, and indeed what is important to all sentient beings.

We know ourselves, if our needs are considered, then our reaction and 'behaviour' will be very different from that if we had our autonomy or choice removed (Schneider, 2012).

If you look at what is going on around the world, and throughout history, one of the biggest causes of conflict is the removal of an individual's choice. If our personal choice, or agency, is removed or restricted, it has a negative effect on our whole being. If choice is continually taken from us, we can develop mental health issues and ultimately a state of 'learned helplessness' (Seligman, 1972).

Learned helplessness occurs when people or animals feel helpless to avoid negative situations.

Such uncontrollable events can significantly debilitate organisms: they produce passivity in the face of trauma, inability to learn that responding is effective, and emotional stress in animals, and possibly depression in man.

(Martin E. P. Seligman, PhD, © 197Z. *Learned Helplessness*, Departments of Psychiatry and Psychology, University of Pennsylvania, Philadelphia, Pennsylvania)

The power of choice is one of the greatest gifts bestowed on man.

(Anon)

In psychology we call choice 'agency'. The ability to act on your own will. If a puppy is to grow up as a freely thinking individual as opposed to a slave of human desires and demands, then it needs to know that it has its own agency. Too much can also be difficult. So, opportunities for agency in a safe environment would be growth experiences for your puppy just like they are for children in kindergarten.

(Maria Paviour, registered occupational psychologist)

We should be offering our puppies an environment that is 'safe' and enables exploration; an environment to enable the development of the puppy's mind and body. This does not necessarily need space but consideration and creativity to offer surroundings that will complement the puppy's physical and mental development. This type of development is also known as enrichment and was originally developed in zoos and aquariums.

PUPPY'S ARRIVAL

When we bring our new puppy home it is one of the most exciting and worrying days: and I am talking about how we feel, and not how your puppy feels! You are bringing a wonderful young puppy into your home. But not just your home, but also into **your** environment, **your** timetable, **your** routine, **your** expectations, and **your** lifestyle, completely into your life.

Whereas your puppy has just been taken away from **their** family, **their** comfort, **their** security, and **their** guardian, and have been placed, often on their own (or with another dog that they have never 'met' before) into a completely new environment (that in so many ways is incompatible for their build and structure), new smells, new everything, *and that was not their choice.*

Consider how traumatic that would be if you were put in the same situation. The opportunity of establishing a positive and sustainable relationship will have greater success if we try to get into the head of the puppy and ponder their needs and perceptions.

Even with the best intentions, we should always consider the consequences of our actions from the puppy's perspective; how our actions will affect their psychological and physical development and how these will impact your life together.

The information about looking after a puppy is readily available. It can be all-encompassing, covering what to feed, how to socialise, exercise, training, leash walk, toilet train, equipment to use, where they should sleep, and so on. There is some good advice around but consider how much is puppy-focused and how much is focused on making it easier for us. Much of this advice is more about the puppy being conditioned so that they can adapt to living in our environment, with our conditions and rules and how much is about their healthy development into adulthood.

The safety of your puppy has got to be of primary concern. Therefore, some advice will involve equipment and practice and is said to be for the puppy's immediate safety. However, we must also consider the puppy's future, to enable them to flourish as adults, especially as some equipment, when used incorrectly, can have huge negative long-term effects on your puppy's well-being. Therefore, even if this equipment may make your life, temporarily, a little easier, it could have significantly bad effects on your puppy's mental and physical development. Think puppy!

This book is to help you 'think puppy', aiding the development of your puppy's body into one that can support a healthy adult body, which will be robust to sustain our modern life, society, and environment.

The recommendations included in this book involve considerations of both your puppy's mind and body, each of which is important for the development of the other; if the body is healthy, the mind will be healthy, if the mind is happy, the body will be healthy.

The areas that will be referred to for a puppy's development will be based on the following:

- Exercise
- Activity
- Equipment
- Environment

These aspects will be considered within the puppy's context through the stages of development and inclusion into our human world, wherever we are in this world.

1 Why is giving correct exercise and activity to your puppy so important?

THE CANINE'S ANATOMICAL HISTORY: FUNCTIONAL ANATOMY AND ETHOLOGY

'Of course he can do it, he's a dog'. I often hear this or similar statements when people are describing their dog's physical capabilities, suggesting that dogs can adapt to whatever challenge we present them with. But how do we know that a dog is coping physically or mentally?

One of the major problems we have is interpreting our dog's levels of comfort or discomfort or even pain. We do not appear to have the ability to observe early signs of physical or mental discomfort; we miss the early signs (Morton, 2005), often only noticing when their behavioural changes are so dramatic that they affect **our** lives. Their stoicism is so great that we often miss problems, such as stress and pain, for weeks, months, and sometimes years.

Another statement I often hear is 'they are lame, but they are not in pain'. This statement could not be further from the truth. Lameness = pain. How worrying is it that our understanding of dogs is based on the belief that being lame or limping is not a result of being uncomfortable at best, and at worst in constant pain.

Human lives, environment, and routine have dramatically changed in just the last 50–60 years. Our lifestyles have altered beyond recognition. Our environment, surroundings, and homes have changed so much and so quickly, that whilst we are happy and able to adapt, we have not given a thought to how these changes have or will impact on our dog's physical and mental health.

In my opinion, these changes in our environment and lifestyle are having radical and negative effects on our dogs' physical and mental condition. In this context, the word 'environment' refers to our houses, our daily routines, work, exercise, and nutrition as well as our life expectations and perceptions.

WHY ARE OUR ENVIRONMENTAL CHANGES HAVING AN IMPACT ON OUR DOGS?

A dog's anatomical form has fundamentally changed very little from the present day to how it looked thousands and thousands of years ago. Throughout history we believe that dogs were living as part of our community; they were living with us, but outside our homesteads. Generally, they were not 'owned' or claimed by an individual, but they lived in their canine community within a human community, a truly symbiotic relationship (Figures 1.1 and 1.2).

DOI: 10.1201/9781003268789-1

Figure 1.1 The Basenji – thought to be the oldest breed still in existence.

Figure 1.2 This picture was drawn 8,000 years ago, depicting dogs of that era. Their basic form and therefore anatomy is akin to many types of breeds now.

Their physical structure appears essentially the same now as it was then. A physical form that has the intrinsic natural ability to traverse substrates that was conducive to the design of their feet and claws, aiding maximum traction.

They lived with other canines, hunted, slept, and played when there was a requirement, and protected the humans of the community, being rewarded with food or the ability to safely scavenge. This functional interrelationship can still be seen in some areas and countries today, like the street dogs we see in various countries of the world (Figure 1.3).

Figure 1.3 Street dogs in India, relaxing together in the centre of the city.

In many urban situations, even up to the post-war era in the UK, dogs lived within households, but were let out in the morning, where they met up with other local dogs and scavenged and played together, forming functioning 'family' groups (Figure 1.4) (Bradshaw, 2011).

For 14,000 years, dogs' lives have had relative environmental and social constants, but in the last 70 years these have changed radically – and we unfairly expect them to adapt, to live in our environment and lifestyles, without us making appropriate adaptations to help them!

(The author understands that these are broad statements and often dogs were tied up outside or had different existences to those that are being discussed. The rationale behind the narrative is to put into perspective how we have expected

Figure 1.4 A typical scene reminiscent of post-war England, with dogs roaming freely to mix, then returning to their homes in the evening.

dogs to radically change within a very small time frame, and not as a global factual statement.)

DOGS' CURRENT LIFESTYLES

Historically, dogs have been selectively bred for specific different functions. Their legs have been bred longer or shorter, their bodies longer, heads bigger, noses flatter, tails taken off (now left on), dew claws removed and, in some breeds, even the arrangement of the limb placement has been positively bred to be altered.

Perhaps we should now breed dogs that *can* accommodate the new most common role, which is housemate, companion, member of a human family group, living within the modern environment and lifestyle.

Recommended reading: What can 'streeties' teach us about companion dogs? | The *IAABC Journal* has a great peer-reviewed article.

Sadly, it appears that within breed selection development, the predominance has not been to breed for the new human environment and lifestyle that dogs now share with us. Motivated by celebrity and fashion (Packer *et al.*, 2017, 2019).

Therefore, many of these 'human' refinements or breed developments have in fact created a dog that is even less able to manage human lifestyle and environmental changes.

THE RAPIDLY EVOLVING WORLD OF THE DOMESTIC DOG

Within living memory, dogs' lives and habits have changed radically, from so many perspectives. One of the biggest changes is how dogs interact and cohabit with us. The domestic environment can prove stressful for companion animals who may respond in different ways depending upon individual differences (Mills *et al.*, 2013). Not just the conditions within our homes, but also how we exercise them, feed them, activities we participate in, the equipment we use, and the restrictions we enforce upon them.

Less than 100 years ago domestic dogs were generally allowed out of their homes in the morning and were let back in at night. Many town and country dogs living in a household were left to roam on their own. They met up with other dogs, scavenged for food, slept, and explored their territory and almost had a 'secret life', but then went home in the evening and ate scraps from the table.

I am most certainly not advocating that we should be resorting back to this type of management. It is now of course, illegal in many countries for dogs to roam. It was also much safer in the days before vehicles dominated our roads in Europe.

What is difficult to reconcile is that we are seeing an epidemic of physical problems[1] with our domesticated dogs, such as osteoarthritis, disc problems, cruciate problems, premature 'ageing', persistent lameness, and also behavioural issues.

Could this be predominantly because dogs are living within an environment that they are not **anatomically designed** to cope with?

To understand why this statement could be affecting our dogs, we need to have a working knowledge of how a dog is structured and what this structure needs to maintain the dog's physical integrity. For this it is helpful to understand how their anatomy works functionally; how the body is organised for efficiency and balance.

TIMELINE

We are expecting dogs to adapt from a lifestyle and environment they have lived in for the last 18,000 years, into a completely new and alien lifestyle and environment (Figure 1.5), without questioning how this may be affecting their health and wellbeing (Figure 1.6).

Table 1.1 demonstrates some of the relevant physical changes that have happened in the canine environment.

We therefore need to play an active part in creating within our environment, a domain where our puppies and dogs can naturally flourish, fulfilling their physical and psychological needs.

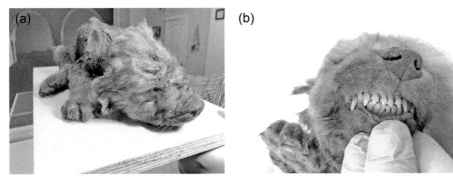

(a) (b)

Figure 1.5 (a, b) 18,000-year-old Russian puppy 'Dogor' found preserved in a layer of permafrost in Siberia. It was found by Russian scientists in 2019. Copyright Dr Sergey Fedorov, North-Eastern Federal University, reprinted with permission.

---

[1] Galen Myotherapy empirical evidence from 20 years of treatment.

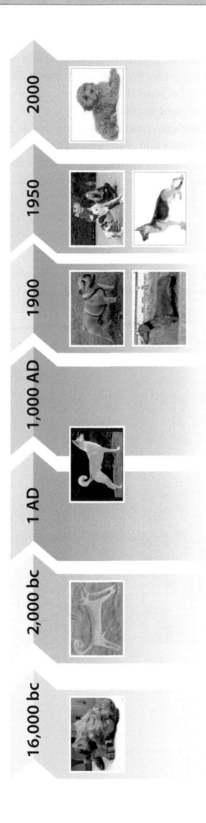

Figure 1.6 Timeline of dog breeds.

Table 1.1   Physical Changes in the Canine Environment Over Time

17,000 BC–mid 20th century (1950)	1970–present day
For at least 18,000 years dogs have lived as follows: 1. Lived outside on natural surfaces. 2. Puppies are born outside. 3. Remain with their littermates. 4. Puppies play and develop with their siblings. 5. They were free to move around freely and explore and understand their environment. 6. Can sleep when they are tired. 7. Have company all the time, lived in canine groups. 8. Can urinate and defecate when they wish. 9. Had autonomy, to eat, sleep, play, fight, establish relationships.	**In the last 50 years dogs have lived as follows:** 1. Live inside on unnatural surfaces. 2. Born inside on unnatural surfaces. 3. Separated from siblings at eight weeks. 4. Play alone or often with dogs larger and stronger than they are. 5. Often caged – limited time and opportunity to explore their environment. 6. Have their sleep routine decided by a human schedule. 7. Often left alone. 8. Can only defecate or urinate when they are allowed out. 9. Their whole day is organised through the human schedule. Points 1–6 can have negative impacts on the physical and psychological development of a puppy. Points 7–9 can have negative psychological impacts on the development of a puppy.

KEY POINTS

- A dog's fundamental structure and anatomy have not changed for the last 18,000 or more years.
- The environment where they now live has changed dramatically and conflicts with their evolved form.
- The breeds and breeding have dramatically changed in the last 100 years, creating an even greater inability to cope with today's environment.

2   *Canine anatomy*

A working knowledge of how a dog is structured

Many of us drive cars and use computers extensively, but do not necessarily understand how they work. However, we all know that if we maintain equipment with moving parts it is going to prolong its life and functionality. For the same reasons (and so many more), we really should have a basic understanding of how dogs and puppies are built, then we can make more informed decisions about their physical welfare.

Following on from the car analogy, I am sure that most people would prefer not to replace their car's engine or a major part because they simply have failed to perform a basic maintenance task, such as checking the oil. Such a simple task, if ignored, can have disastrous consequences. Likewise, we should know our car's limitations; you would not think of driving a road car through deep mud, because it would just get stuck; it is just not designed for those conditions. Or drive a car's engine at full revs, all the time, especially when starting it from cold.

However, many of these motorcar examples can be used as canine comparisons and are virtually duplicated in our management of exercise and activity in our puppies. By understanding how a puppy and adult dog are built and why they are built in the way they are, could possibly prevent a disastrous 'breakdown', which could potentially equate to an impact on their health and well-being.

Given a basic knowledge, it would be great to have a better understanding about how to look after our own puppies' equivalent to 'oil', wheels, and chassis by giving them the correct exercise and activity that will strengthen and develop their structure, producing a robust foundation to support a flexible and healthy body. We really would want to avoid having to replace broken parts.

A SMALL LOOK AT HOW A DOG IS BUILT OR CANINE ANATOMY

The dog or canine's anatomy has not outwardly changed for thousands of years of their recorded existence.

Their musculoskeletal anatomy is constructed in a way that facilitates the potential for incredible mobility, flexibility, and speed, as well as the ability for stealth-like motion (Figures 2.1 and 2.2).

These attributes are only available to a dog when its body can work in balance. In other words, its musculoskeletal system works, in synchronicity, to form efficient, smooth, comfortable movement.

9

DOI: 10.1201/9781003268789-2

Figure 2.1 A dog turning and running – demonstrating a dog's athleticism and freedom of movement, power, and speed.

Figure 2.2 A puppy showing the ability for stealth and stalking, requiring strength and stability.

This *cohesive* physicality develops dynamically as the puppy grows. What this means is that their physical structure requires natural, functional movement to stimulate healthy cellular development of their musculoskeletal structure.

If the physical stimulation is incorrect, at best, their body will not develop fully and will not fulfil its potential. At worst, they will develop an unstable, unbalanced body with unequal loads, and stresses.

Both the dogs in Figure 2.3 are ten years old – but the difference in their posture could be entirely attributed to their environment, activities, and treatment of injuries (Figure 2.3).

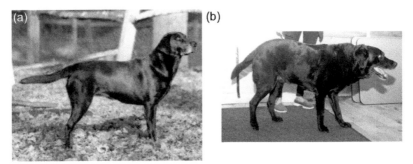

Figure 2.3 Two ten-year-old dogs (a, b) demonstrating the difference in their postures.

The figure shows Maggie at ten years of age (author's dog) versus Jess at ten years of age. Maggie – no balls/slippery floors/not over exercised/injuries treated. Jess – ball throwing/slippery floors/over exercised/not treated (see slippery floors, page 174, and ball throwing, page 193).

THE DOG'S PHYSICAL DEVELOPMENT WILL REFLECT THEIR TYPE OF ACTIVITY AND ENVIRONMENT

Like us, dogs' bodies develop according to use. For example, if an individual just lies on the sofa all day every day, then all of their muscles, soft tissue, and skeleton will adapt to support that lifestyle. If suddenly they do something more active, their structure would not be able to support that change. This is because the body had not been conditioned for a different activity.

To understand this concept, let us initially look at how dogs and puppies are built, in other words, their musculoskeletal anatomy.

We will then take this further and look at how the structure is intended to work, or their functional anatomy.

Once we have studied this, it will clarify why our puppies need to be given additional supporting exercise and activity to stimulate a good physical structure and enable the development of a healthy, mature, adult body.

BASIC ANATOMY

THE SKELETON

The skeleton is the strong part of a dog's structure. It forms a physical protection for some of the main workings of the body, i.e., brain, spinal cord, heart, and lungs (Figure 2.4).

It also accommodates bony attachment points for the soft tissue, such as muscle, tendons, ligaments, and fascia. These soft tissue connections join bone to bone to form joints; convert these 'joined-up' bones into the joints that convert the bones into the levers that with muscle and fascia activation form locomotion.

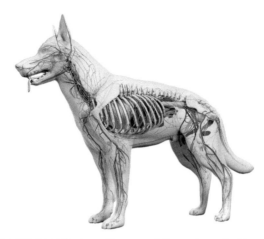

Figure 2.4 Diagram of a skeleton of a dog with some of the anatomy that is supported and protected by the skeleton.

The bones within the skeleton are defined as being dynamic, meaning they will adapt their shape and density according to the load being asserted on them and through them. This is very especially very important when considering the development of the skeleton of a puppy.

When looking at the overall skeleton of the canine, it is almost identical in anatomical make-up as a human (Figure 2.5 and Table 2.1).

When studying the skeleton, it is easier if we divide it into two anatomical parts:

- **The axial skeleton** is the main central body, which is the skull and vertebrae, or spinal column (Figure 2.6).
- **The appendicular skeleton** is the fore and hind limbs, including the shoulder blade or scapula, and also the pelvis plus the axial (Figure 2.6).

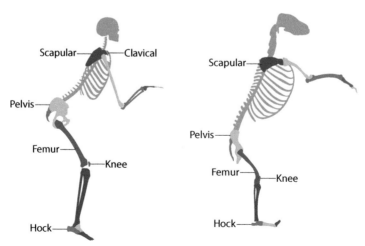

Figure 2.5 A diagram showing the human skeleton on the left is almost identical to the canine on the right.

Table 2.1 Highlighting the Skeletal Differences between Humans and Dogs	
Human structure	**Canine structure**
Number of vertebrae Neck – 7 Thoracic – 12 Lumbar – 5 Sacrum – 5 (fused) Cauda (tail or coccyx or coccygeal vertebrae – 4	Number of vertebrae Neck – 7 Thoracic – 13 Lumbar – 7 Sacrum – 3 (fused) Caudal (tail) – 18–24
Scapula or shoulder blade – lies alongside each other	Scapula or shoulder blade – lies opposite each other
Clavicle or collarbone – stabilises the arm by connecting to the main body	Clavicle or collarbone – no functioning bone present
Thumb – to be used in anatomical 'opposition' (or opposing thumbs), with the fingers, to form a pinch action	Dewclaw – cannot be used in 'opposition' but still has a fundamental use for the dog
Arms – anatomically all the same bone but uses the arms to carry and propel	Forelimbs – anatomically all the same bones as the arm, but walk on their fingers
Legs – anatomically all the same bones but walk on the flat 'plantar' surface of the foot	Hindlimbs – anatomically all the same bones as the leg but walk on their toes

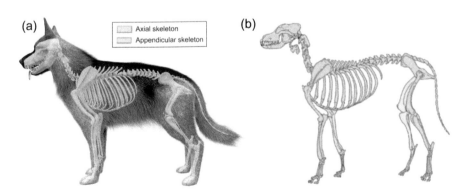

(a) Axial skeleton / Appendicular skeleton   (b)

Figure 2.6 The canine skeleton divided into the axial and the appendicular divisions.

THE AXIAL SKELETON: THE CENTRAL STRUCTURE

THE SKULL
The head or skull is a heavy anatomical structure that contains and protects the dog's brain. The brain is the 'mothership' and the central controller for the whole body, so the structure protecting it has to be robust.

One part of the brain is called the brain stem. The brain stem commands the messages from the brain, and to the brain; it also has a vital role in controlling breathing, wakefulness, blood pressure, heart rate, and swallowing.

The brain responds to the internal environment (how their body feels and what it needs) as well as their external environment (what is happening around them, and to them).

The brain within its different 'compartments' instructs movements and actions, both voluntary and involuntary, as well as receiving and responding to sensory information, and so much more. It is constantly receiving information from the body and feeding back to alter and change the body's responses.

There is a hole (the technical anatomical name is a foramen) at the back of the skull that is the exit and entry point for all the nerves to travel to and from the skull or brain and brainstem.

When nerves travel from the brain, they have to travel via the beginnings of the spinal cord, through the foramen in the skull, to the vertebrae of the neck, or **cervical vertebrae** (see page 19).

The skull also houses the visible sensory apparatus, such as the eyes, ears, nose, and mouth. The skull also supports the jaw.

The jaw

The jaw is built for strength for chewing. It is supported by a very shallow joint within the skull that enables good articulation to facilitate a strong bite. This joint is called the temporomandibular joint (TMJ) (Figure 2.7).

The TMJ, in human terms, can be responsible for many connected disorders and also a great deal of pain. The position of the jaw will also affect the position of the tongue.

The mouth cavity is encased by the jaw; this also cradles the tongue. The tongue is supported by another part of the skeletal system called the hyoid apparatus, which sits near the throat of the dog (the tongue is an anatomical structure that is an important feature of the physical balance of the dog) (Figure 2.8).

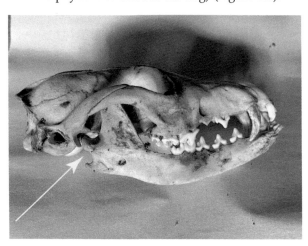

Figure 2.7 Lateral view of the skull showing the position of the jaw joint/attachment.

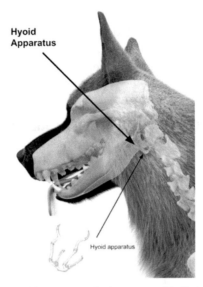

Hyoid
Apparatus

Hyoid apparatus

Figure 2.8 The location of the hyoid apparatus, the bony assembly that supports the tongue.

The tongue is another muscle; it is important for far more than just helping with mastication (chewing). The tongue is essential for balance of the body; according to a study, it can even affect stress perception and management. It can affect lung capacity (Bordoni *et al.*, 2018). The muscles that support the hyoid apparatus are known to help support the posture of the head, which influences the whole movement of the dog.

Therefore, the alignment of the jaw (TMJ) and alignment of the tongue are interrelated.

The skull's foramen is the motorway transport system for the main nerves serving the body. It has two-way traffic, neural messages going from the brain to the body and from the body to the brain (Figure 2.9).

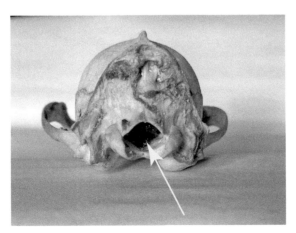

Figure 2.9 View from the caudal aspect of the skull showing the skull foramen magnum.

The joint between the skull and the first cervical (neck) vertebra is called the atlanto-occipital joint, allowing a limited range of movement.

The skull is connected to the first neck vertebrae, or to the rest of the body by soft tissue, consisting of muscle, tendons, ligaments, and fascia that secure the head onto the body.

THE CANINE SPINAL 'COLUMN' OR VERTEBRAL ARRANGEMENT

The word column does not really make sense in the canine world as their vertebral or spinal column is in fact horizontal, and not vertical, as it is with the human.

Interestingly, I have found from my experience of treating humans and canines, the effects of the stresses going through the dog's horizontally stacked vertebrae present equal, if not greater physical challenges for the canine than the vertically stacked human.

It is also my experience that humans and canines share many of the same musculoskeletal issues.

The collective construction and design of the vertebrae, lying alongside each other, form the vertebral or spinal column (Figure 2.10). This column formation forms a long tunnel, like a canal, starting from the first neck or cervical vertebrae, and this canal is continuous all the way along to the sacrum.

VERTEBRA

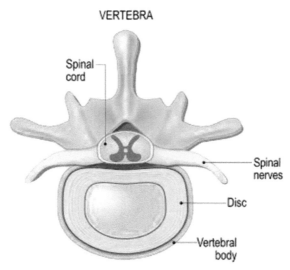

Figure 2.10 A cross-section taken between a vertebrae body, showing where the central nerves arise, and where the intervertebral discs are situated.

This continuous hole through aligned vertebrae or vertebral canal houses the spinal cord that contains the nerves to serve the whole body.

The vertebrae are the bones that are carefully designed stacking blocks. Their external design is knobbly and irregular shaped, intended to support multiple muscle attachments. They also have protrusions that form articulations or joints between each vertebra called facet joints.

Their individual shapes are different for the different areas of the dog. This is to facilitate different movement potentials or provide articulations for protection of the body (thoracic vertebrae, articulates with the ribs) (Figure 2.11).

The muscles that are attached and connecting the vertebral bodies together combine to also maintain the vertebrae's integrity or the correct relative position in relation to their adjoining vertebrae to keep the vertebrae properly aligned.

These intervertebral muscles are highly 'innervated' and carry many nerves within their fibres.

These nerves are very sensitive to a change of positioning within the vertebrae, reporting very efficiently to the brain about any vertebral misalignments, however small. The brain responds by creating a painful spasm within the muscles that acts as acute pain and naturally reduces movement of the canine to protect the alignment against further and potentially catastrophic injury.

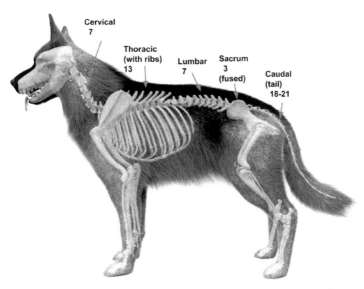

Cervical
7

Thoracic
(with ribs)
13

Lumbar
7

Sacrum
3
(fused)

Caudal
(tail)
18-21

Figure 2.11 A picture of a dog's body showing where all the vertebrae are situated.

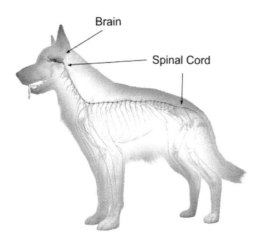

Brain

Spinal Cord

Figure 2.12 The central nervous system with some of the peripheral nerves.

All these individual vertebral bones have intrinsic holes (foramen), to allow an exit and entry point for all the related nerves that serve or are feeding back from that relevant part of the body.

Simply, these nerves will innervate (supply an organ or body part with nerve activity) generally to the closest corresponding location of the body, in relation to the position of the vertebrae (Figure 2.12).

Example: If the puppy's front legs are to move, these nerves will exit through the vertebrae in the lower part of the neck. If the puppy hurts their front paw, the pain neural messages will travel back to the brain through the nerves entering the neck or cervical vertebrae.

> The location of nerve exit points in close proximity to where they innervate is not a rule but it is common. Some nerves originate far from where they innervate, and so is their position. One of these nerves is a cranial nerve called the vagus nerve = the travelling nerve (see vagus nerve, page 67).

THE JOINTS OF THE VERTEBRAE

Facet joints

The vertebral facet joints are fairly similar to other joints in the body; there are two articulating surfaces to facilitate flexing, extending, and rotation of the vertebral bodies.

These 'movement' joints or the facet joints almost look like hands meeting. These structures articulate movement and, critically, maintain the stability and unity between the vertebral joints during the dog's movement.

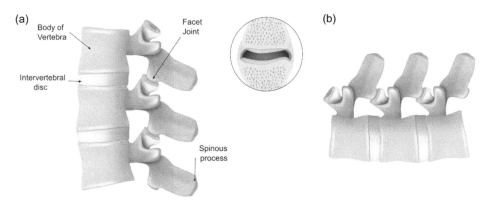

Figure 2.13 The different locations of the intervertebral joints.

The facet joints are synovial joints. They are situated on the top (dorsal) aspect of the canine's vertebral bodies or bones (these are the ones that are 'adjusted' by therapists such as chiropractors and osteopaths) (Figure 2.13).

Intervertebral joints

However, this vertebral construction has additional vital functions; the intervertebral joints facilitate the absorption of concussion felt through the dog's body. The 'bounce' absorbing joints are between each vertebra and are situated on the ventral underneath the vertebral bodies. This modified joint structure is where the infamous intervertebral discs are located.

These intervertebral discs consist of modified cartilage that is designed to absorb forces and concussions received through the body. These discs look like a flattened cushion or sac containing viscous, gelatinous fluid that is designed to accommodate and absorb normal forces made and received by the body, protecting the central neurological system or spinal cord.

If any of these structures receive consistent, excessive loads or force, especially asymmetric forces (from one or other side of the body), concussion, or compression, these structures or 'discs' can be forced out of their retaining membrane, and like a water-filled balloon, can be squeezed through gaps and end up pressing on the nerves arising from the corresponding vertebra.

If the compression is severe, then the bulging disc can physically restrict the peripheral nerves arising from the spinal column. In human terms, this is often called a 'slipped disc'. In canines, it can be called a bulging or a herniated disc (Figure 2.14).

It could be said that there are two vertebral joints:

- The facet joint for vertebral movement.
- The intervertebral joint to absorb bounce/concussion.

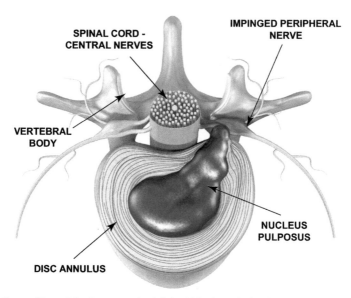

SPINAL CORD -
CENTRAL NERVES

IMPINGED PERIPHERAL
NERVE

VERTEBRAL
BODY

NUCLEUS
PULPOSUS

DISC ANNULUS

Figure 2.14 The position of the intervertebral disk within the spinal column when there is herniated or Intervertebral disc disease.

THE NECK OR THE CERVICAL REGION (ANATOMICALLY TERMED C1–C7)

(The Latin word cervix means neck, so this word can be used in other areas of anatomy to describe a 'neck', i.e., the neck of the uterus is called the cervix.)

The neck (cervical) vertebrae are the conduit for all the nerves leaving (efferent nerves) the brain and arriving (afferent nerves) back to the brain from and to the body (Figure 2.15).

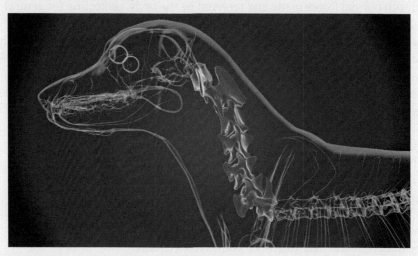

Figure 2.15 Demonstrating where the neck vertebrae, or anatomically called the cervical vertebrae, are situated in the dog.

A CRITICAL NEURAL HIGHWAY

The first two cervical vertebrae have modifications that are peculiar to each of them, to allow for greater mobility to facilitate the 'yes' and 'no' actions.

The joint between the skull and the first cervical vertebra is called the atlan-to-occipital joint. The atlas (named after the god Atlas supporting the heavens) is the first neck or cervical vertebrae. The atlanto-occipital joint facilitates flexion or nodding of the head, so the atlas is often called the **'yes'** vertebra (Figures 2.16 and 2.17).

The second vertebra, called the axis, the 'no' vertebrae. This is an extraordinary structure that is overlooked from a mobility perspective. It has a unique protuberance that lies within the vertebral body called a 'dens' (Latin for tooth) and allows for the sideways action of the 'no', giving stability through this action.

The axis also supports, by way of a strong attachment, a cord-like structure called the nuchal ligament (Figures 2.18 and 2.19). This is a tough modified ligament that runs continuously from the top of the axis vertebra; this is situated above the cervical vertebrae of the neck, then attaches to the second thoracic vertebra.

This modified ligament then changes its name to the supraspinous ligament, forming strong physical attachments with the top (dorsal) aspect of the vertebrae until the last of the lumbar.

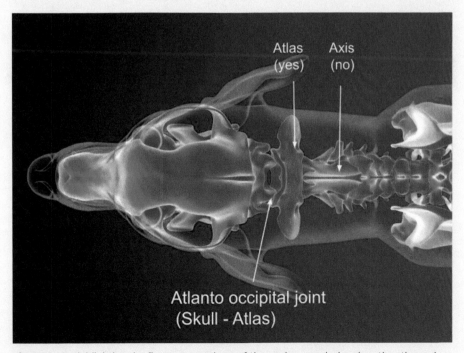

Figure 2.16 Highlighting the first two vertebrae of the neck or cervical region, the atlas and the axis. Atlas with the wings for strong muscle attachment.

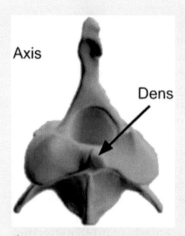

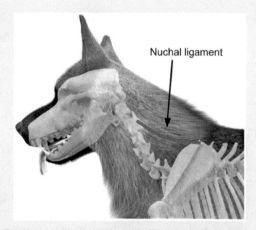

Figure 2.17 A view of the axis with its unique dens.

Figure 2.18 The position of the nuchal ligament, connecting from the axis to the first thoracic vertebrae.

So much of the strength and stability of the neck vertebrae, and in fact total balance of the body, is derived from the functionality of the nuchal ligament. It is also vital for enabling the lifting action of the dog's head off the ground, using the ligament's intrinsic recoil action to pull the head up, rather than relying on muscle strength, which would be highly fatiguing.

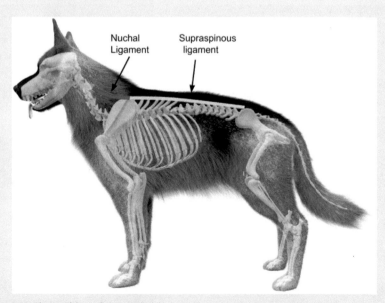

Figure 2.19 The position of the nuchal ligament along with the supraspinous ligament.

Interesting fact: Humans and horses have their nuchal ligament attached to the nuchal process of the skull, but dogs' nuchal ligament does not reach the skull but the first vertebrae. This gives their head more flexibility (so they can reach all the way to their genitals to maintain cleanliness).

This is important because if structures are more flexible, they will be less stable.

THORACIC VERTEBRAE
The rib area or the thoracic region (anatomically termed T1–T13)

There are 13 thoracic vertebrae in total. All of them articulate with a pair of ribs, one rib on each side of each vertebra (Figure 2.20).

It is important to know that ribs are connected through articulating joints with the thoracic vertebrae and therefore are mobile and not rigid fixtures. The ribs (or anatomically named costals) need to have a movable joint to enable the expansion of the lungs during breathing. The ribs also provide bony protection for the heart and lungs.

The section of the ribs that lies at the ventral aspect or underneath the dog's chest is actually cartilage and not bone. This also allows for more flexibility within the

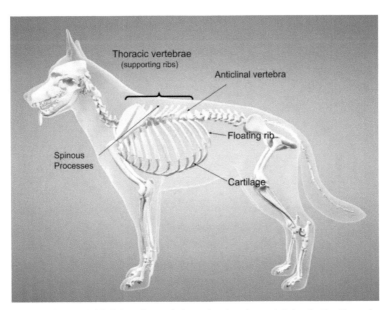

Figure 2.20 The skeleton highlighting the main bony landmarks pertaining to the thoracic vertebrae.

whole thoracic cage. These cartilaginous ends of the ribs form joints at the base of the dog's body and are called the sternum.

The sternum starts at the chest, between the forelimbs of the dog. The breastbone, or manubrium, which is the start of the sternum, can be felt quite easily. The other end of the sternum, which is situated at the other rear or caudal aspect of the ribs, is called the xiphoid process (pronounced ziffe-void).

The canine has a floating rib, which is generally the last rib. A floating rib means it is not attached to the sternum, only to the thirteenth vertebrae. Sometimes, these can protrude on the puppy or dog and look really uncomfortable, but they are generally totally normal.

The thoracic vertebrae derive their stability from the costals or ribs that form a cavity containing many of the body's vital organs. However, this stability also translates as a lack of flexibility through the region, it can rotate slightly, but cannot flex and extend as the cervical or lumbar region can.

The thoracic vertebrae subtly change direction, and at the point of change, it is called the anticlinal junction (Figures 2.21 and 2.22).

Interesting fact: This region has the capability of allowing the spine to hyperextend, absorbing the momentum when they are stopping coming down a hill or at speed.

Figure 2.21 Picture showing the type of movement when the vertebrae absorb this downwards load at the anticlinal point when the dog changes direction (taken from DVD *Tongue to Tail: The Integrated Movement of the Dog.* Julia Robertson and Elisabeth Pope).

Anticlinal vertebra
between T10-11
*(anticlinal meaning an
arch)*

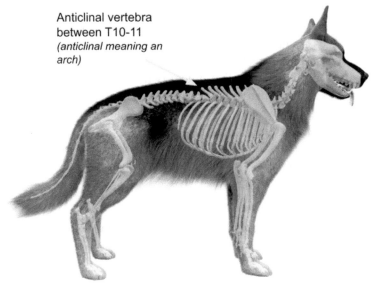

Figure 2.22 The position of the anticlinal vertebrae within the context of the thoracic vertebrae.

THE LOWER BACK: LUMBAR REGION (ANATOMICALLY TERMED L1–L7)

There are seven lumbar vertebrae in total. This area is often referred to as the 'lower back'. The construction of the lumbar vertebrae is intended to have mobility that enables flexion, extension, and rotation (Figure 2.23).

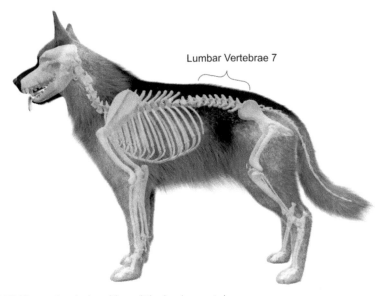

Lumbar Vertebrae 7

Figure 2.23 The anatomical position of the lumbar vertebrae.

The flexibility through this region is a distinctive factor to the canine and is what gives them superior speed and flexibility, especially when compared with other quadrupeds such as the horse. This flexibility is coupled with the action of the sacroiliac joint (see sacrum, page 27).

The lumbar vertebrae maintain their integrity through having a good muscle and soft tissue support matrix. There is a complexity of abdominal (stomach muscles), and muscles that are interwoven with fascia, called obliques. These obliques form a criss-cross type of sling around this abdominal region.

The abdominals and obliques are important because they initiate movement and aid the stability of the lower back or lumbar region. These muscles are part of the core muscular system.

> **Interesting fact:** Due to flexibility within this region, in dogs, as in people, this region is generally regarded as the most vulnerable to injury or referred pain.

The more flexible a region is, the more vulnerable during locomotion and often being the first to have reduced flexibility.

Comparing the equine movement to the canine, the fundamental differences are the lack of flexibility of the horse compared with the dog through the sacroiliac and lumbar regions. A clear visual representation of that is the angle of the pelvis through flexion (retraction) and full extension (protraction) whilst galloping.

RETRACTION OF THE LIMBS

See Figures 2.24 and 2.25.

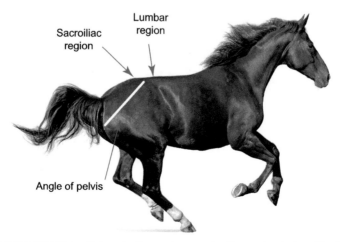

Sacroiliac region

Lumbar region

Angle of pelvis

Figure 2.24 A horse galloping – demonstrating and comparing the anatomical positioning of the pelvic structures.

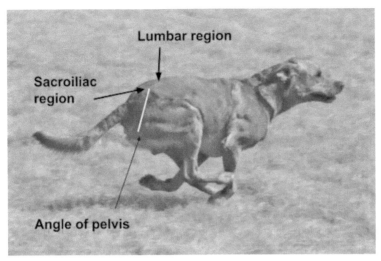

Figure 2.25 A dog galloping – demonstrating and comparing the anatomical positioning of the pelvic structures.

PROTRACTION OF THE LIMBS

See Figures 2.26 and 2.27.

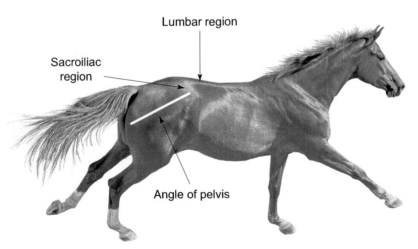

Figure 2.26 A horse galloping – demonstrating and comparing the anatomical positioning of the pelvic structures.

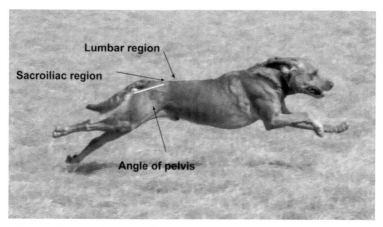

Figure 2.27 A dog galloping – demonstrating and comparing the anatomical positioning of the pelvic structures.

THE SACRUM

The sacrum is different from the other vertebrae; it comprises three vertebral bodies, but rather than articulating with each, they are fused together.

These vertebrae literally have a pivotal role, they are positioned on the shallow articulating surface of the pelvis. This joint (called the sacroiliac joint), allows for the 'rocking' action between the sacrum and the pelvis, which is required to transfer the movement and drive from the hindquarters and pelvis to the rest of the body (Figure 2.28).

Figure 2.28 A dog galloping – demonstrating where the sacroiliac region is situated.

Interesting fact: These fused vertebrae form the joint that connects the vertebral column onto the pelvis. and therefore, to the 'engine'. This connection allows a pivoting action through the joint (sacroiliac joint) that allows the dog's hind limbs, when running, to reach far underneath their body. This gives them a huge range of movement, which, coupled with the mobility of the lumbar region, form a big difference in the range of movement of a dog compared with a horse.

THE CAUDAL VERTEBRAE, OTHERWISE KNOWN AS THE TAIL

As humans we do not have a functional tail, but for the canine, it is an important structure for both movement and communication.

The dog's tail is incredibly important from both a physical and a behavioural perspective. We all know about reading a dog's tail to know if they are happy or sad, but there is so much more in between those two emotions that are expressed through their tails. A wagging tail doesn't always mean happy; a tucked under tail can also mean so much more than just sad or worried.

However, from a physical perspective, the tail can act like a fifth leg. If a dog's body is healthy and working well, when running in a straight line, the tail tends to be quite relaxed. However, when cornering, or turning, it helps to counterbalance the dog. It also helps to slow the dog down from running or going downhill (Figures 2.29–2.31).

Figure 2.29 Balancing along a structure is aided using the tail.

Figure 2.30 Jumping down is aided by the tail.

Figures 2.31 This series of pictures took place in ten seconds, and within that time, the tail was fully engaged to assist with the arrival, stopping, and lifting of the body, then leaving the area.

The actions of the tail often happen quicker than the human eye can compute, so can therefore be discounted as far as their importance to the function of the canine.

The less balanced or painful a dog is within its body, the more it uses its tail to aid motivation. In some extreme cases of pelvic weakness, a dog can lose part of or a high degree of the function of its tail.

> **Interesting fact:** This dog is ten years old, but often dogs of this age have lost most of the function of their tail due to postural changes (Figure 2.31).

> **Interesting fact:** The tail can be used like a parachute, to slow a dog down, to assist poor function in its legs, or to help stabilise the dog. From a behavioural perspective, it is a vital tool for communicating with other species, especially other canines.

THE APPENDICULAR SKELETON: ARMS AND LEGS, AKA THE WOBBLY BITS!

If the axial skeleton is the central structure of the dog's body, then the appendicular skeleton comprises the appendages, or the limbs.

The dog's appendicular skeleton, or the limbs, are anatomically a direct equivalent to our limbs.

The dog's forelimbs are almost the same construction as our arms, and their hindlimbs are almost identical to our legs.

The colours of both pictures show the comparison between the skeletal arrangement of both the human and the canine.

A big difference is that a dog stands on what is anatomically comparable to our fingers on their front legs and their toes with their hind legs.

THE FORELIMB: LOWER LIMB CONSTRUCTION

THE TOES OR PHALANGES

A dog stands on its toes like a human holds a claw-like grip. Their nails are directed towards the ground and the 'palm' of their paw is seated firmly on the ground (Figure 2.32).

The claws have different functions from that of human nails; the dog's claw can retract (draw in) and protract (draw out) depending on the surface or action (Figure 2.33).

NB: We have often seen when a dog is on a slippery surface and their instinct is to protract their claws, causing even less traction and therefore more sliding.

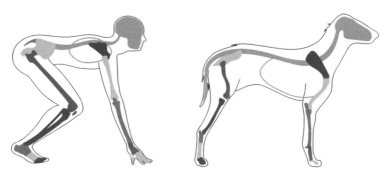

Figure 2.32 Demonstrating the comparable anatomical bony placement between the human and the dog, with the human taking on the quadruped stance.

Protracted claws for gripping

Figure 2.33 The paws and claws are vital for more than stability and traction when moving; they are required to secure and hold food to enable chewing.

THE WRIST OR CARPUS

When a dog is static or standing still, its wrist is not 'planting' or in touch with the ground. However, when a dog is running, stopping, or jumping the wrist hyperextends to absorb the forces through the forelimb. This helps to protect the rest of the leg as well as the physical shock travelling into the main body and vertebrae.

The wrist or carpus hyperextends primarily through the loads being directed through the tendons situated in the carpus joining the feet to the leg. These tendons act as a cushion but also help with some recoil, giving the leg more impetus and spring into the next stride.

Stopping
See Figure 2.34.

Figure 2.34 A running dog, stopping quickly to perhaps change direction, using the absorbing hyperextension of their carpal or wrist joints to aid the absorption load.

Running
See Figure 2.35.

Figure 2.35 A running dog demonstrating the hyperextension of the wrist during the landing phase.

Jumping down
See Figure 2.36.

Figure 2.36 A dog jumping out of a car; a dog naturally hyperextends their carpus or wrist when they are stopping, turning, or landing.

A dog's wrist is often misnamed as their 'knee' but anatomically that is completely incorrect.

THE FORELIMB FOOT

Anatomically, the canine forelimb foot and the human hand are almost identical. The main difference is in how we have adapted their use. The canine foot has to absorb downward force, whereas the human hand is much more dextrous and can perform different actions and fine motor skills with the fingers of phalanges.

DOGS' FRONT PAD

The other anatomical similarity on a dog's forelimb is their dewclaws. These are in the same anatomical position as our thumb (Figures 2.37 and 2.38). Their dewclaws do not have the ability to produce the 'opposition' or opposing action against other digits like humans and other primates can (Figure 2.39).

The dog uses that dewclaw in the same way as their other claws; and the dewclaw protracts and retracts simultaneously as the other claws. They use the dewclaw to grip an object, maintaining traction, especially when turning, and to form grip when they are scaling banks, fencing, or other obstacles (Figures 2.40–2.43).

Interesting fact: When the dog's wrist hyperextends, the dewclaw naturally comes in contact with the ground, allowing a natural functional application.

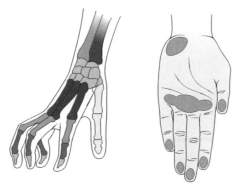

Figure 2.37 The bones of the human hand plus the load points, where the hand would take the load of the ground if it were compared with a canine.

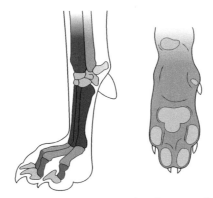

Figure 2.38 The canine paw both anatomically and showing the areas where the dog loads when moving.

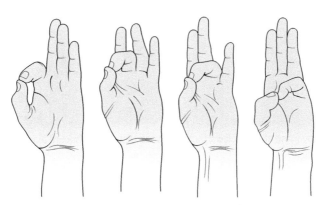

Figure 2.39 Thumb opposition is something that a dog cannot replicate.

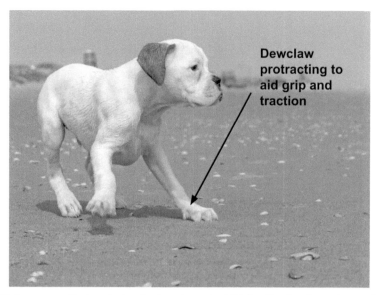

Figure 2.40 A puppy on the beach showing how important it is for the dewclaw to be used for traction and grip during activity.

Figure 2.41 A puppy using all their claws, including their dewclaw, to secure a bone so that they can chew and eat it easily.

The dew claw protracted to aid stability whilst turning

Figure 2.42 A dog running and turning suddenly, using their dewclaw to help with the turn by adding medial stability. (Collie pictures taken from the DVD Tongue to Tail: The Integrated Movement of the Dog – Julia Robertson and Elisabeth Pope.)

Figure 2.43 A close-up of the positioning of the dewclaw as it gains traction in the ground.

THE FORELIMBS: UPPER LIMB CONSTRUCTION

All the joints of their forelimb articulate or bend in the same direction, or plane as a human's arm (Figure 2.44).

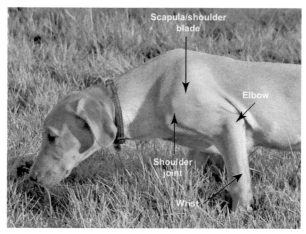

Figure 2.44 In this wonderful picture of a young and fit dog's shoulder and forelimb, some of the comparable anatomy of the human arm is clear to be seen.

THE CLAVICLE OR COLLARBONE

The dog has one vital difference to the human, which is in the upper forelimb. They do not have a similarly functioning collar bone, or clavicle. Some dogs do have a small cartilaginous structure that lies within a muscle group. This 'clavicle' is referred to as vestigial, or a past purpose, but we do not really know if this was ever a functioning clavicle.

The lack of a clavicle in the dog's structure is because they do not have the need. Their natural functions and actions do not require that type of stability.

The canine requires more flexibility within the scapula and upper limb so that they can perform fast coursing or zig-zagging action, creating a secure but flexible movement at speed.

Likewise, they require the ability to 'stalk', which involves almost dropping their body between their scapulae, or shoulder blades.

Importantly, dogs do not need to extend their arms laterally or carry heavy loads on their 'arms'. Both actions are important for humans, and therefore need the stability of a collarbone or clavicle (Figure 2.45).

Humans need the bony stability of their 'arms' or forelimbs so that they can perform tasks such as carrying heavy weights and suspending the load through their shoulders (Figure 2.46).

The clavicle allows humans to safely open our arms widely or perform an abduction. This is different for dogs who just have muscle supporting this action. This is why any form of extreme abduction movement is so detrimental to their

Figure 2.45 Dogs do not need their 'arms' to carry heavy weights like humans seen here carrying two heavy bags of shopping.

Figure 2.46 Unlike dogs, humans can abduct (open) their arms safely.

musculoskeletal stability. One such action is slipping on slippery flooring within their environment (see slippery floors, page 174).

In humans, the clavicle is attached from the manubrium (sternum) to the acromion process of the scapula. These connections would not work with the canine's anatomical layout, having their scapulae opposite each other, as opposed to alongside in humans (Figures 2.47–2.49).

Figure 2.47 The approximate position of where a clavicle would be positioned on a dog's body.

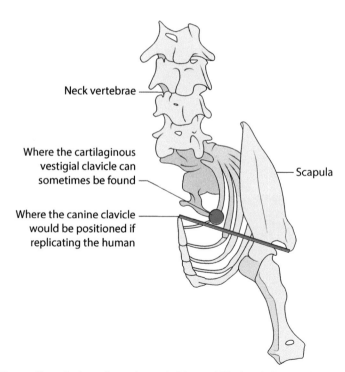

Neck vertebrae

Where the cartilaginous vestigial clavicle can sometimes be found

Scapula

Where the canine clavicle would be positioned if replicating the human

Figure 2.48 The position of where the canine's clavicle would be in relation to a human.

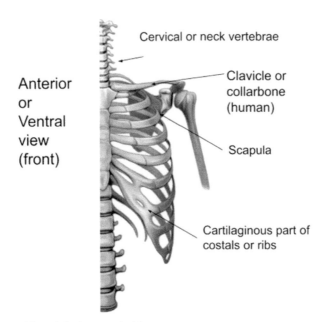

Cervical or neck vertebrae

Anterior
or
Ventral
view
(front)

Clavicle or
collarbone
(human)

Scapula

Cartilaginous part of
costals or ribs

Figure 2.49 The position of the human clavicle.

The lack of a functioning clavicle means they must rely on balanced muscle attachments for stability within the forelimb.

Anatomically this means that for the dog there is **no bony attachment** from the forelimb (arm) to the rest of their body. The dog's forelimbs are stabilised and connected to the rest of their body by muscle and associated soft tissue.

Dogs need the flexibility through their shoulders to 'stalk' and move slowly. This ability is often lost through muscle compensatory changes when they move into maturity (Figures 2.50 and 2.51).

Figures 2.50 Pictures showing a puppy getting into a stalking pose, the shoulder 'sling' that is a soft tissue support system, allowing the body to dip through the shoulders.

Figure 2.51 A puppy demonstrating a 'play bow' to possibly initiate play.

ELBOW AND ELBOW JOINT

Another important joint is the elbow joint. This joint is a hinge joint that has restricted movement to flexion and extension (bending and straightening). This joint also receives much of the concussion and impact and bears the weight of the puppy's proportionately heavy front end during all types of movement (Figures 2.52–2.56).

Figure 2.52 A young dog running fast and demonstrating how the elbow takes a huge kinetic load when the puppy is active.

Figure 2.53 A dog standing still and watching, demonstrating how the elbow takes static load when the puppy is still.

Figure 2.54 A puppy jumping or falling down will add impact to the elbow joint.

Figure 2.55 A young dog moving slowly will exert a huge load through their elbow.

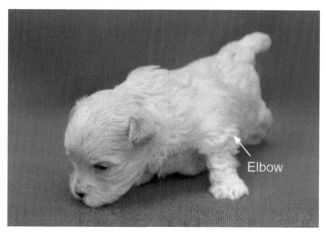

Figure 2.56 Every time a puppy stands from a sit or lying down position, it will drive a force through their elbow, especially if they are getting up on a slippery surface (see puppy sit, pages 177 and 201).

THE SHOULDER AND SHOULDER JOINT

This is a ball and socket joint and has a bigger range of movement but a smaller range than the human equivalent. With that potential for movement, it is equally important for it to be stable enough to accept huge loads and changes of movement. To facilitate this, the joint has a complex of ligaments, maintaining its structure but also muscles and tendon placement to provide good stability. There are strong movement muscles that provide the flexing and extending of the arm or forelimb (Figure 2.57).

Figure 2.57 A puppy running and extending their shoulder and elbow.

THE SCAPULA

The scapula for the dog and human is very similar, but the one very big difference is their alignment with each other. On the human they are positioned side by side; on the canine they are positioned opposite each other (Figures 2.58 and 2.59).

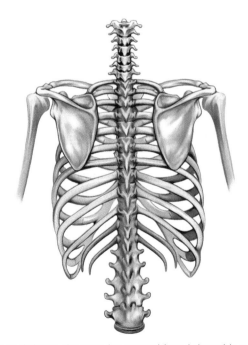

Figure 2.58 On the human skeleton, the scapulae are positioned alongside each other.

45

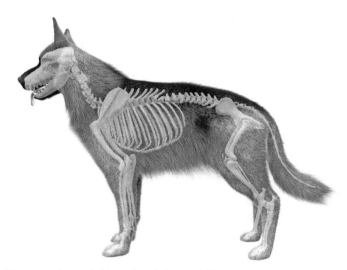

Figure 2.59 When you place a visible canine skeleton within a dog's body, it can be clearly seen that the scapulae are positioned opposite each other, as opposed to a human's scapulae that are alongside each other.

The scapula is attached to the body of the dog by muscle and fascial attachments. This structure is the major attachment point for the forelimb (appendicular) to the main skeleton structure (axial). It is also one of the major components of the shoulder sling that is often referred to when talking about the forelimbs of a dog. The 'sling' pertains to the supporting structures of both the bony structures as well as the soft tissue, muscle, fascia, tendons, and ligaments.

MOVEMENT POTENTIAL OF THE CANINE'S FORELIMB

The canine's front limb has limited lateral (sideways) and medial movements. Anatomically the leg can only move away and towards the body's midline by about 45 degrees, which is significantly less than humans (see Figures 2.46 and 2.60).

We must remember that the dog walks on its arms; this is important when we consider the physical differences between human and canine structures and the impact of how they perform all forms of locomotion. This takes on a very different context when you look at how the neck is also involved in all the impaction of forelimb action (see neck, page 195).

Figure 2.60 A dog demonstrating lateral movement through the forelimbs. It is a critical aspect to the movement, stability, and flexibility of the dog – but this action has a limited natural range due to their anatomy and shoulder construction.

THE HINDLIMBS

THE HINDLIMBS: LOWER LIMB CONSTRUCTION

The dog's hind legs are anatomically identical to our legs. As with the forelimb, the dog is walking on the equivalent of our toes (Figure 2.61). In some breeds, there is a 'fifth' toe or hind dewclaw; sometimes they are double dewclaws (Figure 2.62).

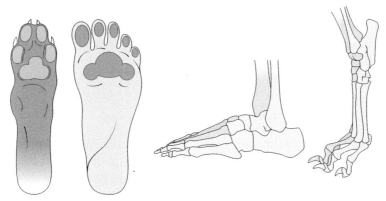

Figure 2.61 The comparison between the human and canine foot, the bone anatomy (on the right) and plantar surface of the canine compared to the human foot.

47

Figure 2.62 A puppy St Bernard showing the presence of hind limb dewclaws.

These in most breeds are vestigial, or without direct function, but there are still some that have bony attachments. They are thought to be the remaining anatomy from when dogs used to climb or require additional traction, and these would then act as a functional dewclaw or extra toe. Especially for mountain breeds, this would have been imperative.

THE HOCK

Their hock is anatomically the same as our ankle. But unlike us, the dog's 'plantar' surface (or our foot), is not constantly in touch with the ground (Figures 2.63). The only time the dog's plantar surface or foot touches the ground is when they run, stop, jump up, or take off from a standstill (Figures 2.64, 2.65, and 2.66).

This mechanism is another spring-loaded apparatus within the dog's anatomy that allows the downwards load or force to be converted into a recoil action.

As with the nuchal ligament (see neck, page 20) to assist with the head lifting their head, or the hyperextension of the wrist or carpus, this structure helps minimise muscle usage and maximises energy usage.

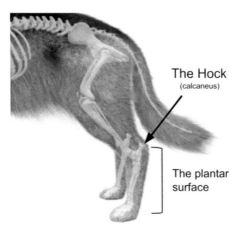

Figure 2.63 Part of the skeleton that on the dog is called the hock (also called the calcaneus, one of the bones of the anatomical dog's foot).

Figure 2.64 A puppy demonstrating how they use the plantar surface of their foot and 'leg' to gain more traction through increased surface contact when they wish to jump up.

The flexibility given through the hock is facilitated by a complexity of interweaved and isolated connecting soft tissue, comprising tendons, fascial, and ligaments. A particularly large tendon attachment is the Achilles tendon, which attaches at the point of the dog's hock.

Using the plantar surface of hind limb

Figure 2.65 Dogs playing and demonstrating the importance of enhanced drive and traction using their full plantar surface.

Achilles tendon

Hock

Using full plantar surface of hind limb

Figure 2.66 Dogs playing and demonstrating the importance of the flexibility within their lower hindlimb for added strength, traction, and receipt of physical load.

Interesting fact: Often people wonder why their dogs' have grass stains on the surface of their legs, this explains that dogs use their lower limb or 'foot' in a similar way to humans.

Interesting fact: Two of the three hamstring muscles, which are the large and powerful muscles that push the dog forwards, have long tendinous attachments onto the region of the hock.

THE KNEE OR STIFLE JOINT

The knee or the stifle joint of the dogs is remarkably interesting. It is of similar construction to the human knee.

The knee is under a huge amount of load, plus downward and rotational pressure, due to much of the dog's momentum being carried and transferred through their stifle.

The design of the stifle is highly intricate, comprising of overlapping soft tissue, almost containing the joint as a layered parcel. The deepest part of this 'wrapping' is integral ligaments called cruciate ligaments. They lie deep within the centre of the joint. They are called cruciate, meaning cross-shaped, because they connect the joint diagonally (Figure 2.67).

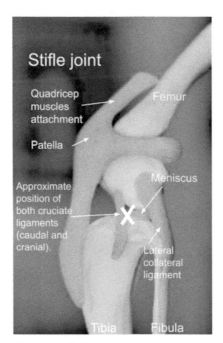

Figure 2.67 The stifle joint, with many soft tissue layers removed, showing the internal structures of the cruciate ligaments, and more external located lateral collateral ligaments (one on each side) and the meniscus, situated like cushions within the joint.

There are also two 'pads' called the meniscus, they add cushioning on either side of the joint to help protect the structures from excessive concussion.

Then two ligaments laterally hold the femur and tibia together, called the collateral ligaments. Their operating construction is akin to a pulley and lever system, the bones (femur and tibia) being the levers and the patella (kneecap) acting as a pulley.

The whole mechanism supported by this intricate arrangement of soft tissue, tendons, and ligaments collaboratively form a very strong joint.

The stifle should only flex and extend and should be prevented from rotational, or twisting, forces. Its integrity relies on the foot placement during movement being correctly and evenly placed with good stability through the whole limb.

The thigh bone or femur, or one part of the 'lever', has a specially adapted groove (trochlear groove) which receives and maintains the tendon location of the quadricep muscles in their true anatomical position (the action of the quadricep group of four muscles is to extend the knee, a kicking action).

The quadricep muscle tendons are then attached to the patella (the pulley) then further soft tissue (ligament) is attached to the tibia (the other lever). When the stifle flexes, the patella aids the flow of the tendons through the anatomical groove.

The combination of the levers, the groove, and the patella (or pulley) facilitates a smooth sliding action to effect good muscle and joint action. These combined tendons use this groove to maintain their position in relation to the femur, keeping the action of the knee true. The patella is key to this 'lever and pulley' system (Figures 2.68–2.70).

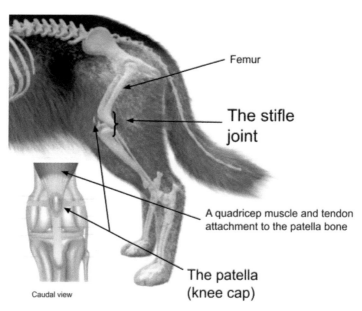

Femur

The stifle joint

A quadricep muscle and tendon attachment to the patella bone

The patella (knee cap)

Caudal view

Figure 2.68 The position of the stifle or true knee of the canine.

The stifle

Figure 2.69 A flexed stifle or knee on a running adolescent.

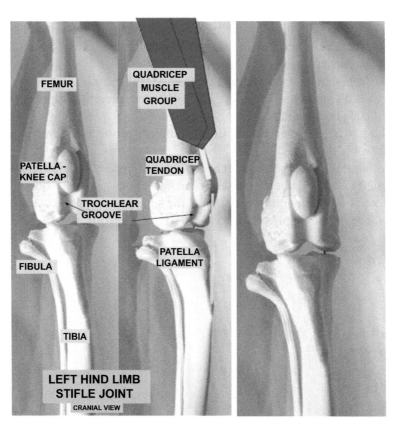

FEMUR

QUADRICEP MUSCLE GROUP

PATELLA - KNEE CAP

QUADRICEP TENDON

TROCHLEAR GROOVE

FIBULA

PATELLA LIGAMENT

TIBIA

LEFT HIND LIMB STIFLE JOINT

CRANIAL VIEW

Figure 2.70 The anatomical structure of the stifle, labelled and not labelled showing the mechanism of the trochlear groove of the femur.

Interesting fact: Due to its design and action, the stifle joint is vulnerable; it relies majorly on maintaining its correct function from having good hip stability, as well as good conformation or structure. Without either or both, the knee mechanism will be compromised.

If the trochlear groove is not sufficiently deep to maintain the leverage of the muscles and tendons through the joint, a dog could be suffering from a conformational condition called subluxating patellas. This condition could cause what is colloquially known as the 'Jack Russell hop'.

THE HIP JOINT

The hind limb is connected to the pelvis and therefore the main body, via the hip joint.

As in the human, the hip is a ball and socket joint. The ball is on the end of the thigh bone or femur and the socket (acetabulum) is within the pelvis. This joint is also called the coxofemoral joint.

The hip joint has the ability to move in multiple planes of movement. It can move the hind limb away from the body (abduction), take the leg towards and also under the body (adduction), rotate (circumduction), flex, and extend. There are multiple potential movement planes or directions.

This joint relies on its stability primarily from a complexity of various size muscles attaching to and around the hip to keep the ball seated correctly within its socket, and to facilitate good flexibility.

The joint also has an integral ligament that directly connects the ball of the femur into the socket of the pelvis.

It is the 'fit' of the ball and socket that defines the integrity of the joint. If the fit is not good, the dog could be diagnosed as having various degrees of hip dysplasia. The fault can lie with either the ball (head of the femur) or the socket being of an incorrect shape or size (Figures 2.71 and 2.72).

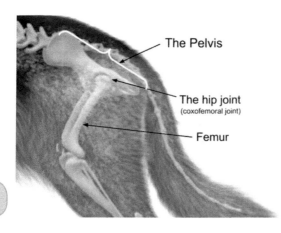

The Pelvis

The hip joint
(coxofemoral joint)

Femur

Figure 2.71 The exact location of the hindlimb skeletal placement, highlighting the hip, or anatomically known as the coxofemoral joint.

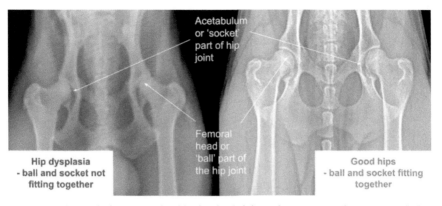

Figure 2.72 A radiograph demonstrating hip dysplastic joints of a mature canine compared with the good hips of a mature canine.

MUSCLES, FASCIA, AND OTHER SOFT TISSUE

- Muscles connect to bones by tendons (Figure 2.73).
- Bones are connected to each other by ligaments to form joints.
- Fascia – 'one muscle six hundred pockets', Lowri Davies: BVSc, Dip ACVSMR, MRCVS, Cert Vet Acup.

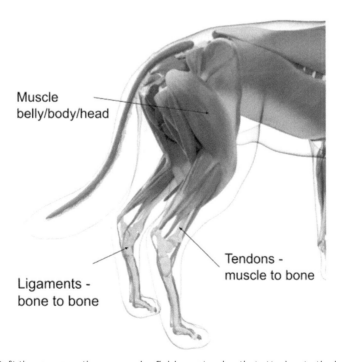

Figure 2.73 Soft tissue connections – muscles finish as a tendon that attaches to the bone. Ligaments join bone to bone.

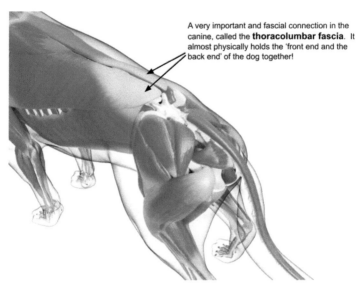

A very important and fascial connection in the canine, called the **thoracolumbar fascia**. It almost physically holds the 'front end and the back end' of the dog together!

Figure 2.74 The thoracolumbar fascia, a thickened facial connection over the lumbar region of the dog.

A concise description of fascia – it is a type of soft tissue that is contiguous throughout the whole body. It is present in various densities, strengths, and appearances, and forms connections; it physically connects muscles to form synergistic strength; and the coordination of muscle action and muscle patterning (see page 95).

The muscle latissimus dorsi attaches to the medial aspect of the humerus (forearm) then across the body connecting to the pelvic region through the thoracolumbar fascia (Figure 2.74). This is part of the facial network acting also as a widened thickened tendinous connection (aponeurosis).

MUSCLES

Muscles and fascia are defined as 'soft tissue'. These components work cooperatively to support the dog's structure or skeleton and to enable locomotion.

It is the muscles and fascia that hold our skeleton in our recognisable shape and posture.

Each individual muscle contains millions of specialised muscle cells. Each of these microscopic cells reacts to neurological stimuli to form action potentials.

To create muscle cell activity takes a highly complex cellular interconnection. It requires conductivity, or electrical stimuli from the nerves, to trigger calcium exchanges that stimulate the sliding mechanism or contraction of each single muscle cell.

How muscles work (simplified)

Muscle cells are capable of contraction – **creating an action**.

Muscle cells are capable of active relaxation – **allowing an action** (Figures 2.75 and 2.76).

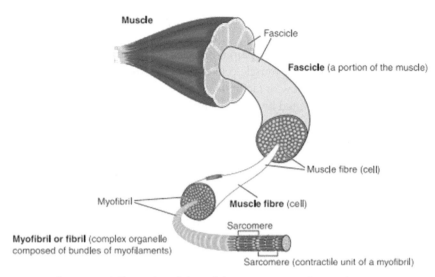

Muscle

Fascicle

Fascicle (a portion of the muscle)

Muscle fibre (cell)

Myofibril

Muscle fibre (cell)

Sarcomere

Myofibril or fibril (complex organelle
composed of bundles of myofilaments)

Sarcomere (contractile unit of a myofibril)

Figure 2.75 A diagrammatic illustration of the cellular arrangement of a muscle.

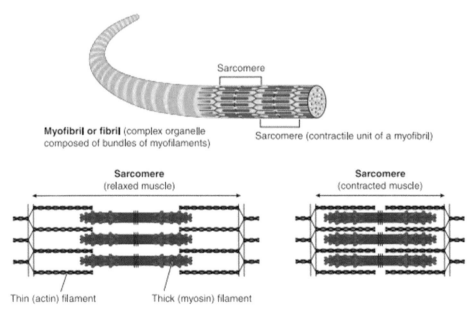

Sarcomere

Myofibril or fibril (complex organelle
composed of bundles of myofilaments)

Sarcomere (contractile unit of a myofibril)

Sarcomere
(relaxed muscle)

Sarcomere
(contracted muscle)

Thin (actin) filament

Thick (myosin) filament

Figure 2.76 The cellular action of the muscle cell when it contracts and how this then is translated into a muscle-looking bulkier, or hypertonic.

The contraction is formed by neurological 'action potential' stimulating muscle cells sliding together, this creates a shorter length within the whole muscle that is called 'muscle contraction'. It also gives a visual appearance of increased muscle bulk.

This involves targeted muscle cells and fibres contracting, enabling a joint to articulate within its own range. With the elbow joint, it will be either flexing (closing the angle between the two bones) or extending the joint (increasing the angle between the two bones).

The contraction and relaxation of paired muscles demonstrate the 'bulking' of the contracted muscles through the muscle cells sliding together. Plus, the importance of the 'relaxing' muscle to allow or facilitate an action.

Muscles react or are activated by appropriate neurological impulses for the action required. They 'actively' relax by not being innervated.

Within the limbs, muscles work in pairs. This means there is a muscle on one side of the bone connected to an articulating joint. The muscle on the opposing side of the bone acts as its 'pair'. Both these muscles are connected to the same joint, but in different positions, which act as different leverage points to enable different actions within the joint.

A muscles contracts (agonist) – articulating the joint (flexing or extending) (Figures 2.77, 2.78, and 2.79).

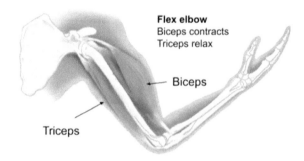

Figure 2.77 The human arm and the muscle recruitment required to flex or bend the elbow.

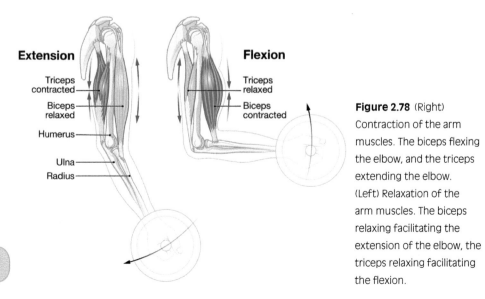

Figure 2.78 (Right) Contraction of the arm muscles. The biceps flexing the elbow, and the triceps extending the elbow. (Left) Relaxation of the arm muscles. The biceps relaxing facilitating the extension of the elbow, the triceps relaxing facilitating the flexion.

Figure 2.79 Muscles working functionally, demonstrating their appearance when contracting and relaxing.

The opposite muscles (antagonist) actively relax, allowing the movement; by doing this, the antagonist must, in effect, become longer than its normal anatomical length.

It is almost like a push and pull action, **but muscles can only pull** by contracting, and the 'push' is not a push, but the allowing of an action through **active relaxation** (Figure 2.80).

- Muscle contraction = nerve stimulus to muscle fibre.
- Muscle actively relaxing = **no** nerve stimulus to muscle fibre (Figure 2.78).

Flexing the elbow   Extending the hip

Figure 2.80 For actions to be efficient and balanced both in flexion and extension of the joints, the paired muscles must be balanced and have equal range and strength.

Interesting fact AND important point:

- Contraction of muscles is vital for all actions; even muscles that are partially damaged can still manage to contract.
- However, a damaged muscle has reduced capability to actively relax (due to the cellular sliding mechanism being damaged).
- When the antagonist muscle is damaged, this reduces how much contraction can be applied by the agonist, therefore inhibiting the range of movement.
- **This inhibition of muscle action is one of the major causes of reduced mobility.**

Within the body, muscles also work as functional pairs, but not in such an obvious way as with the limbs.

The pairing of body muscles is more about body and limb attachment balance.

Muscles over the top of the dog's body must have equal potential range and strength as those in the corresponding position under the body to create and maintain balance for limb and body support and movement.

Within all movement and activity, all muscles must be in balance to work to create a full non-impinged range of movement (Figures 2.81 and 2.82).

Muscles creating the balance between the shoulder and thigh of the dog are critical for balance and stability, but this balance can be damaged and compromised by overloading and stressing from many activities and environments.

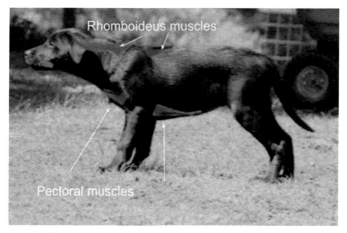

Figure 2.81 Demonstrating how muscles on the dog's body are intended to work in balance. Showing the m. rhomboideus over the top line and the m. pectorals of the chest.

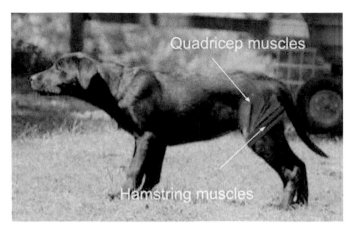

Figure 2.82 Demonstrating how muscles on the dog's body are intended to work in balance. Showing the quadricep group and one of the three hamstrings of the hindlimb.

ROLES OF MUSCLES

Muscles that perform movement are called skeletal muscles or voluntary muscles. Other muscles within the body are the heart, muscle lining within the gut, and artery walls. These are called involuntary muscles and have a different cellular construction.

The voluntary muscles are then divided into two distinct groups:

- Deep muscles – are situated closest to the skeleton and tend to be shorter. Their role is to **support and provide stability for the joints**.
- Superficial muscles – located just under the skin. They predominantly provide the **movement of the dog** by converting the limbs and joints into working levers.

Both deep and superficial muscles are equally important. Without the deep muscles, the skeleton would be wobbly and unstable. It would not safely be able to perform well, slowly, at speed, or in the multidirectional movement for which a dog is anatomically designed (Figure 2.83).

Deep muscles keep the limbs well connected to the body so that the dog can safely use the superficial muscles to their full potential in the same way as wheels must be properly connected to the chassis of the car, so the car is secure and can be safely driven (Figure 2.84).

Good muscle action needs:

- Unimpeded muscle contraction and active relaxation.
- Paired limb muscles working in balance.
- Paired body muscles working in balance.
- Deep stabilising muscles providing sufficient stability within the skeleton, so the superficial moving muscles have a secure framework to perform actions.

The hip and pelvic region has got to be strong and stable to manage the forces

Figure 2.83 An adolescent demonstrating how power is derived from a strong, stable yet flexible body.

Figure 2.84 If the wheels of a car are not secure, they will wear and then come off!

FASCIA

This is a fascinating area of anatomy with its relationship to movement.

For a long time, this white tissue that was interlaced throughout the body was rather ignored by anatomists and surgeons alike. However, over years of science and therapy, it is now viewed with greater respect and understanding. So much so that some are suggesting that fascia should be classed as a **body system**, in the same way as circulatory, digestive, and neurological systems, for example.

Fascia could be thought of as being akin to a body's internal 'wrapping' in that it physically connects all the parts of our body, to provide support, communication, and physical coordination.

Fascia is a type of soft tissue that is also connective tissue in the most real terms; it connects all the body together.

There are various thicknesses of fascia all over the body. The thickest is strong enough to carry the weight of a dog; the thinnest can be as delicate as a spider web.

Some of the functions of fascia:

- Fascia connects muscles in functional lines (front to back/left to right/side to side), therefore connecting actions.
- Fascia joins the body's individual moving parts to enable collective balance and coordination.
- Fascia acts like a sausage skin over muscles, enabling containment of force and strength.
- Fascia is like a cradle to support internal organs.
- Fascia connects the whole body, connecting the thinnest fascia to the toughest.

Fascia has been often described as a form of tensegrity (which means floating compression, components under compression inside a network of continuous tension).

The principle of 'tensegrity' first precisely described the relationship between connective tissues, muscles, and the skeleton.

The word 'tensegrity' was first used in the 1960s by the artist, inventor, and mathematician R. Buckminster Fuller, who coined the term 'tensegrity' from the two words 'tensional and integrity'.

It has now been popularised by Tom Myers (author of *Anatomy Trains*) within the context of fascia, and how it describes the intricate relationship between the connective tissues of fascia and muscles (myofascial) and the skeleton.

It could almost be compared in design to the Kurilpa Bridge in Brisbane, Australia, which is the world's first tensegrity designed bridge (Figure 2.85).

Figure 2.85 Kurilpa Bridge, Brisbane, demonstrating 'tensegrity' within a physical structure. The structure contains what are known as 'compression bars' inside a network of continuous cabling that forms a physical compression that supports the bridge.

This 'self-supporting' bridge is held together by equal compressional tension of the cables, which holds and maintains the poles in position, but still allowing movement.

These cables could be compared to the dog's connecting fascia and muscles (or myofascial connections, myo = muscle), supporting the skeleton in position. Like a body, if one of those cables is damaged, it will affect the whole integrity of the bridge. It is all connected in the same way as it is in the anatomical structure of a body.

It could almost be thought of as like an inflated balloon supporting the skeleton. If you push onto one side of the balloon, it affects the other side.

Fascia is a complex concept to grapple with, so let's use a very basic analogy of air in a balloon. With a balloon, the pressure required to maintain the balloon's inflated structure and shape comes from the pressurised air within it (Figures 2.86 and 2.87).

Figure 2.86 The effects on a balloon, how depressurising one side loads the other side. Therefore, any impact will affect the whole.

Figure 2.87 How different rates of a balloon from a fully inflated, supported structure to a deflated unsupported structure relate to how fascia that has integral equal pressure can maintain the body's integral strength and integrity.

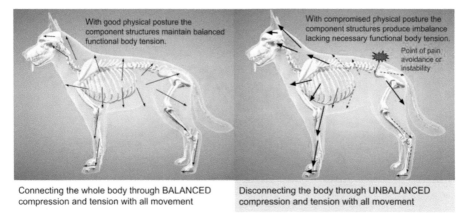

Figure 2.88 A skeleton demonstrating how fascia maintains the connective integrity of the body, but if damaged or compromised, it will lose its whole balance.

If that balloon loses internal pressure, then the strength and integrity of the structure as a whole is lost because it relies on its internal pressure forming the external force to maintain the shape and structure of the balloon.

If this integral 'body' pressure is interrupted by an injury or an instability, then the integrity of the whole is altered, and even loading or pressure is redistributed (see the following diagrams) (Figure 2.88).

Fundamentally, the denser fascia is connected intrinsically with muscle, the less dense fascia is involved with supporting internal structures; all these are contiguous connections, but specifically those of muscle-to-muscle fascial connections form functional lines to connect movement patterns.

This means that fascia joins all the muscle movements together, supporting the different actions and connecting the front of the dog to the back! It also acts as a method of stabilising movement and joining all the individual muscle actions together to form smooth, synergistically stronger coordination actions.

FASCIA: FUNCTIONAL LINES

These fascial lines form like a conduit through the body, which connect muscles and fascia, to enable different actions, together with giving the whole body a supported yet flexible structure.

This structure is also physically efficient, these connections facilitate combined muscle activity, forming strong muscle patterns, as opposed to individual muscle action, through combining the different muscles that provide strength, stability, and speed (Figures 2.89 and 2.90).

If facial lines are damaged or interrupted by muscle malfunction or damage, it has detrimental effects on mobility.

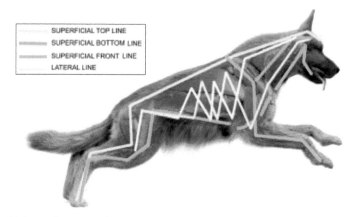

Figure 2.89 This is a rudimentary diagram of how some of these physical fascial lines connect the body, incorporating the muscles to form 'functional lines', so movements can be 'joined-up'.

Figure 2.90 The comparison between the canine and the human of one of the functional fascial lines, indicating the importance of how the head and the hindquarters are fascially connected.

The big difference between the fascial lines of a canine to that of a human, or even an equine, would be the inclusion of the tail.

The tail of the canine is vital for all forms of movement, from directional changes, stopping, assisting movement, and supporting the body. The tail therefore would have to be included within this system. **Look at every picture in this book of a dog moving and look at how they are using their tails** (see page 28).

THE NERVOUS SYSTEM: THE BODY'S ELECTRICAL SYSTEM

The neurological system is complex and works universally throughout the body. Some of its many responsibilities are:

- Receiving and delivering messages.
- Influencing change and the body's balance.

- Creating movement and body awareness.
- Exciting the body – creating stress.
- Relaxing the body.
- Altering changes within the body systems.

The brain, or the body's 'mothership', is in total control. Control of our whole body. It is constantly being fed from stimuli produced from our external environment (the world around us), as well as our internal environment (our own body's internal systems).

Example: The brain is recording dangers or safe situations, from sensory organs such as eyes, ears, skin, and nose; as well as recording if we are tired, hungry, or stressed.

This constant recording and feeding back to and from the brain are what keeps us alive, well, and safe.

The neurological system manages a messaging system that reports to all our body systems, such as our heart, muscles, and digestive system (Figure 2.91).

The brain and its allied neural or nerve pathways have control over so many of our body processes. For instance, it can stimulate the pituitary gland, which is a tiny gland at the base of the brain. The pituitary gland or the 'master gland' is the hormonal mothership of the body, creating complex hormonal adjustments, aiding the maintenance of internal body balance such as monitoring our oxygen levels, salt level, and so on. When a body is in balance it is called homeostasis.

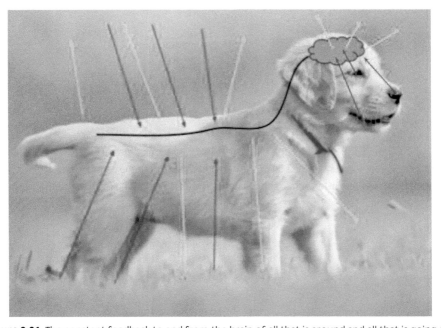

Figure 2.91 The constant feedback to and from the brain of all that is around and all that is going on, inside and outside of the body.

The nervous system comprises the:

- Central nervous system (CNS).
- Peripheral nervous system (PNS).
- Cranial nerves.

The CNS is the brain and the nerves travelling down the spinal cord.

The PNS is divided into two distinct sections, the somatic nervous system and the autonomic nervous system. The somatic nervous system controls the motor nerves, and they do exactly as their name suggests: They provide a 'motor' for the body, and they stimulate muscle movement. These nerves originate in the brain, travel down the vertebrae (CNS), then exit the vertebrae, serving each corresponding part of the body (PNS).

The PNS nerves are big enough to be easily visible under the skin. They behave almost like a hosepipe carrying water. If the nerves (or hosepipe) have not been impinged, the messages (or water) flow to the target. But like a hosepipe, if it is kinked, the flow of messages (or water) to the muscle will be reduced (Figure 2.92).

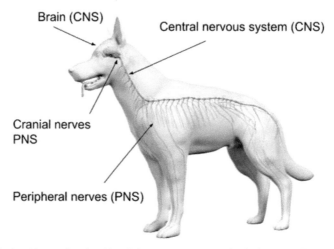

Figure 2.92 The location under the skin of the nervous system, both the central nervous system, the brain and the nerves within vertebrae, and the peripheral, the ones arising from the central nerves from the vertebrae.

The cranial nerves. Even though these nerves are situated in the head, by the jaw, and close to the brain, they are technically classed as peripheral nerves.

There are 12 pairs of cranial nerves (one for each side of the body), and they innervate and feedback detailed information from the eyes, ears, nose, face, and mouth (including structures such as the tongue), as well as posture, balance, and voice.

There is one cranial nerve that is the most fascinating and furthest travelling, the vagus nerve. The vagus nerve takes its name from the Latin meaning 'wandering'. The nerve is extremely long and innervates many different parts of the body's digestive tract, heart, and other internal organs. The vagus nerve responds to the dog's perceived environment, whether they feel safe or under threat. It is a key component to the body's response system for calming.

SYMPATHETIC NERVOUS SYSTEM: FIGHT AND FLIGHT

Another function of the nervous system is something called the 'autonomic nervous system'; it is more commonly known as our 'fight or flight' and 'rest and digest' zones or preparing the state of bodily readiness. **To be ready to run or to be ready to eat.**

This system works on a super-fast short circuit mechanism, with a dedicated nerve supply, travelling directly to the organs or systems that need to be activated to produce an almost instant response.

This specialised function is dedicated to ensuring the body is ready for the next activity, whether it is as diverse as running for your life or settling down to a relaxed meal.

The two zones are called:

Sympathetic nervous system = fight or flight.

Parasympathetic nervous system = rest and digest.

FIGHT OR FLIGHT

See Figure 2.93.

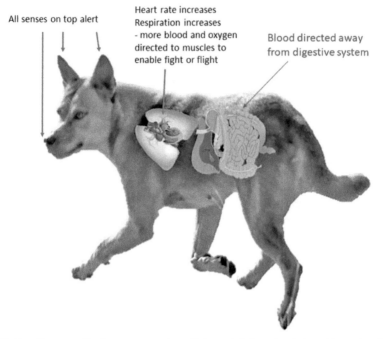

All senses on top alert

Heart rate increases
Respiration increases
- more blood and oxygen
directed to muscles to
enable fight or flight

Blood directed away
from digestive system

Figure 2.93 How the sympathetic nervous system (fight and flight) stimulation affects the different body systems, including the circulatory and digestive systems.

PARASYMPATHETIC NERVOUS SYSTEM (REST AND DIGEST)

See Figure 2.94.

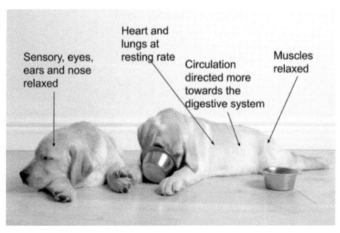

Sensory, eyes,
ears and nose
relaxed

Heart and
lungs at
resting rate

Circulation
directed more
towards the
digestive system

Muscles
relaxed

Figure 2.94 How the parasympathetic nervous system (rest and digest) affects the body, calming and therefore facilitating good digestion.

IMPORTANT NOTE

As the body is connected physically and physiologically, rehabilitation of the body should accommodate this by working to reconnect the whole, not individual parts.

Likewise, connection of the body, such as puppy development should also involve the whole.

KEY POINTS

- Anatomy is a huge and complex subject, but understanding the very fundamentals, it will help understand how your puppy works.
- By understanding a little about how your puppy 'works', it will be easier to understand the information and the reason for the activities to develop them.
- By understanding a little about how a puppy is built will also help to understand how types of exercise and activities can be detrimental to their development and future well-being.

3 *The physical constituents of a 'joined-up puppy'*

Four cellular components are fundamentally involved within the process of developing the 'joined-up puppy'. They are components of the growing puppy that we can influence, both positively and negatively.

Therefore, these are going to act as a foundation to a programme to develop our 'joined-up puppy' (Figure 3.1):

- Brain (and neural pathways)
- Bone
- Muscle
- Fat

The rate of a puppy's growth (Figure 3.2) is extremely rapid, which makes it even more important to assist the establishment of your puppy's physical components.

STAGE 1: BRAIN (AND NERVES OR NEURAL PATHWAYS) DEVELOPMENT

THE BRAIN

An approved socialisation programme, as previously mentioned, is essential as part of the puppy's development. The idea is to gently introduce a puppy to its environment, including both living and inanimate features. These programmes need to start as soon as a new puppy arrives at their home, and **it has to be conducted before the puppy is 17 weeks old** (Figure 3.3).

A puppy should be given the opportunity to be introduced **carefully and gradually** to their environment and as many environments and situations as possible (Figure 3.4).

DOI: 10.1201/9781003268789-3

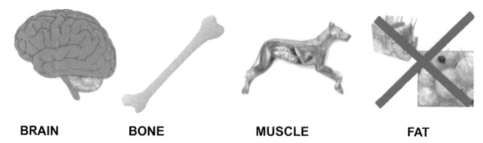

BRAIN   **BONE**   **MUSCLE**   **FAT**

Figure 3.1 The overarching physiological areas of growth with their organisation or progression of the relative growth of brain, bone, and muscle, and the unwanted excessive fat deposits.

Figure 3.2 Wellington Neil, eight weeks to 12 months. (a) Eight weeks. (b) 12 weeks (three months). (c) 16 weeks (four months). (d) 20 weeks (five months). (e) 24 weeks (six months). (f) 28 weeks (seven months). (g) 12 months.

Figure 3.3 A picture from a puppy's perspective. We must try and understand a puppy's view on their new environment from their perspective, getting down to their level, seeing the world with all its 'new' information.

IMPORTANT INFORMATION

Socialisation should be done with professional guidance from an approved canine behaviourist from an organisation. It is vital to gain the correct amount of stimulation for your puppy, together with supportive advice, to understand how to manage situations, how long to expose a puppy, and how to support their post-learning care.

Objects and situations may be common to you, but to your puppy, they could be terrifying, and they may perceive them as life-threatening. They need to learn that they are safe with you and that they can have the time, space, and appropriate environment for them to recover and process positively what they have seen, heard, and smelt (for more details about socialisation, see recommendations, pages 71 and 73).

THE PERIPHERAL NERVES

Developing the brain and peripheral nervous system, specifically the somatic nervous system, is important when the puppy is young. The somatic nervous system delivers innervation, or nerve pathways, to the skin and muscle (the neurological system, see page 64). Ensuring that these nerves are being stimulated is an important part of your puppy's initial physical development so that the puppy has good 'electricity' or nerves to all parts of its body and musculature.

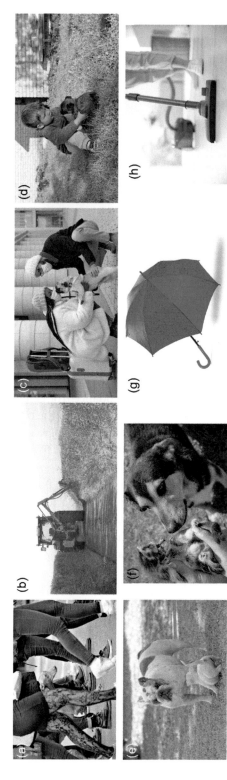

Figure 3.4 Just a few situational and environmental examples of what to introduce to your puppy. (a) Town – noises, people, and vehicles. (b) Country – noises and vehicles. (c) Adults – with masks and beards. (d) Children – all ages. (e) Other dogs – big and small. (f) Other animals – different species. (g) Objects – occasional. (h) Objects – everyday.

The best way to visualise the process of developing neural pathways is to imagine a field of long grass, with a couple of established footpaths crossing through the middle. These paths only go in one direction, from a to b. But this field could lead to so many other potential destinations if different paths were formed. These 'tracks' would then stay with the puppy into maturity, it would then be up to the type of activities performed as to whether it would remain in the brain, this is called 'competitive plasticity' of the brain, or simply 'use it or lose it' (Figure 3.5).

This is how your puppy's nerve development could be envisaged. The paths running through the field are, in effect, the somatic nerves, currently going to one place. However, there are so many other potential paths that could be developed, and these additional paths, or neural pathways, would improve the potential of your puppy's neural connections. By stimulating all the potential pathways, you will ensure that their foundation muscles, their power muscles, in fact, their whole muscular system in their body is joined up and connected. This is a vital part of developing your 'joined-up puppy'.

To ensure your puppy is creating as many 'paths' as possible, they have to be introduced to different stimuli. These stimuli are created by encouraging your puppy to use different movement patterns, directions, and speeds.

Figure 3.5 A field showing travelled pathways, conceptually depicting the importance of developing as many neural pathways as possible in the developing puppy.

Applying this varied activity will assist the stimulation of the structural muscles as well as the power muscles, which will then be activated by the appropriate neural supply.

The muscle and neural linkage will ensure the puppy's body is 'joined-up' to create a stable, robust structure. A puppy with good stability within its whole structure will be more flexible, mobile, and sturdy. They will also have a greater ability to cope with physical activity and its associated rigours.

Somatic nerve stimulation, and muscle activation and development, are interwoven. The two are inextricably linked.

Good somatic neurological connections also enhance:

- **Spatial awareness**–awareness of where their body is in relation to other objects.
- **Proprioception**–a puppy's ability to be bodily aware within their environment, of both their movements and actions, without requiring conscious thought.
- **Coordination**–how the puppy uses different muscle combinations to create different movements and the ability to isolate muscle groups to create more complex movements (Figure 3.6).

A puppy with good spatial awareness and proprioception will suffer less physical injury, due to them being able to cope within their physical environment.

It is also important to stimulate your puppy's cerebral development to help them cope with issues in their life, including through problem-solving and 'cause and effect' (see brain games, page 155; Figure 5.34).

Figure 3.6 An example of great coordination, spatial awareness, and proprioception – a young dog running, dipping, and turning under a branch.

OLFACTORY NERVE DEVELOPMENT (THE SMELLING NERVES!)

There are other parts of the brain and associated nerves, including cranial nerves that many of the socialisation programmes do not necessarily fully target. One of these specific areas is the development of a puppy's scenting capabilities.

A dog's superpower is its nose. A dog's nose is 100,000 times more sensitive than ours. Maybe, because we cannot empathise with the importance of this 'superpower', we perhaps forget how important it is for a puppy's and mature dog's emotional, and also physical, fulfilment (Figure 3.7).

The nose and scenting ability, or olfactory system, like all senses, has dedicated nerves receiving information and transporting them back to the brain, then from the brain to a body response.

The canine also has potentially 40% more brain dedicated to receiving scents, however, as with all neural pathways, it needs stimulation.

The puppy's nose still needs stimulation in the same way as every other working part of its body. The puppy is born with all the neural pathways, but they still need appropriate arousal to fully activate these connections; in other words, like muscle neural connection, the nose or scenting skill also needs stimulation. **It is vital for a puppy to develop these natural skills**, as their scenting will also serve to increase their understanding of their world and environment (Figure 3.8). With practice, it will give them the ability to interpret and discriminate between different scents and smells, allowing them a deeper knowledge and comprehension of their world.

If puppies are not encouraged to use their nose, they may not reach their full scenting potential, or they may even 'forget' to use it. This could put your puppy at such a disadvantage and potentially compromise their fulfilment.

Figure 3.7 The puppy's nose is highly efficient, at least 10,000 to 100,00 times more than a human's nose.

Figure 3.8 As soon as your puppy arrives you can start scenting games and exploration. This is Marge Tyrrell, a 12-week-old Westie-poo, using her nose to explore, and therefore, have a greater understanding of her environment.

If a puppy has not had a good start with a responsible breeder, it may not have had the initial engagement or opportunity to use its nose. Likewise, if a puppy lives in a sanitised human environment, the smells can be so overpowering they may choose not to use their olfactory skill, due to them being so overwhelming.

As previously mentioned, what is good for a puppy psychologically, is also good physically. Using their nose is one of the best examples of this. When your puppy puts their nose to the ground and sniffs, this natural body position, static or moving, will promote good neural development of the important structural muscles. **It is an important activity for developing a stable structure**.

Encouraging your puppy to use their nose, and the ability to refine this 'superpower', will help them to find fulfilment, calm, fun, purpose, and also a level of autonomy (see games and activities, page 132).

BONE, SKELETAL DEVELOPMENT

Within this section, the discussion of bone development concentrates on **long bones**, which are found within the limbs or appendicular skeleton (see page 30).

As the puppy grows and gets taller, the long bones lengthen. The bones lengthen by a complex process known as ossification. Ossification occurs within the 'growth plates', these growth plates are situated at both ends of every 'long bone'.

This rate of ossification coincides with the growth of the puppy, which of course varies hugely in different breeds. Some breeds or breed types have been known to take up to 18 months for their long bones to fully ossify, or mature.

It is often considered that the larger the breed, the longer it takes for the growth plates to ossify, but this cannot be guaranteed; some smaller dogs have been shown to have late ossification maturity.

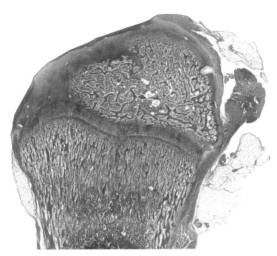

Figure 3.9 A really clear diagram showing the end of a long bone and the position of the epiphyseal plate or growth plate, when the bone is still in the growth cartilaginous and vulnerable phase (as in the figure) and when it has ossified and become more robust bone tissue.

Each puppy has their own rate of development and unless we take clinical images, there is no real way of knowing how long that may be (Figure 3.9).

The 'growth plates' that are in a puppy's long bones are soft cartilage until they have ossified into hard bone. As the puppy grows and develops, ossification converts this soft cartilaginous tissue, or cartilage, into bone.

WHY IS THIS IMPORTANT WHEN WE ARE GROWING A PUPPY?

During growth, until maturity, the long bones on a puppy are vulnerable, as the growth plates are near the articulating surface (joints) and are under many mixed loading forces due to different movements and activities taken by the puppy.

During this time, the type of exercise and activity should be appropriate to assist the correct structural development of these bones and joints.

For a long bone to fulfil optimum strength, it is really important for it to receive appropriate 'forces' so that the bone can respond by developing strong functional lines within the long bone. These functional lines will be the strongest part of the bones that are required to support the activity conducted by the puppy and mature adult.

If a puppy has ongoing restricted mobility, even the restricted free movement around their environment/home, this will have a negative impact on bone strength and correct functional strength lines within these long bones.

Too much of the wrong exercise and too little of the correct exercise will be detrimental to the puppy's structural development.

To repeat – **too little movement can be as detrimental as too much movement!**

It is often cited throughout this book, look to nature. What would a puppy do if left within a natural environment? They would run and play, then sleep, play for a

Figure 3.10 More damage can be done to a puppy by restricting them. For healthy development, they need to be able to 'free run' to strengthen their bones and their minds (see free running, page 128; see crating, page 180).

few minutes, then sleep, walk, and play hunt, scavenge, sleep, and so on. Puppies need appropriate exercise coupled with appropriate rest.

If puppies are not given enough 'free running', as they would in a natural environment, their bones will not receive the appropriate forces required for good dense bone development (Figure 3.10).

However, there is a fine balance at this stage, because during this period, excessive and repetitive patterns of exercise should also be avoided. A puppy should be allowed to naturally free run; this is when a puppy is running without encouragement, lure, or being pursued by a larger or adult dog.

Bones continue to adapt to loads and stresses as the puppy develops into maturity. Long bones require natural loading for them to create or maintain cellular mass.

However, these cells can also negatively modify if an unbalanced load or loads are asserted. A bone will alter its areas of strength to adapt to where the load is more demanding.

What does this mean in a dog's life? To maintain the optimum skeletal strength, the loads of activity conducted by the puppy, adolescent, and mature dog must be as balanced as possible. Therefore, we really need to be working to establish and maintain our puppy's symmetrical structure

MUSCLE

This section covers individual muscles and muscle group activation, stimulating the correct muscles to perform an action.

This section looks at developing the 'kinetic chain' or connecting all the individual muscles to the fascial connections to form correct movement or movement chains that will form strong, flexible, and balanced and connected movement patterns.

Correct muscle development in a puppy is critical because it is the muscles that have influence over the symmetry of the skeleton.

The emphasis is on the word **correct** muscle development. Not all muscles are the same, different muscles have quite different roles. For a balanced dog, all these muscles should work effectively in their different roles (see muscles and their roles, page 85).

Developing your puppy's correct musculature will set them up for life!

If puppies were living in a natural and outside environment, and growing up in their family group, we would not have to actively consider how to correctly develop their musculature. But they don't: they live in a **human environment**, so we **do** need to assist their development (Figures 3.11 and 3.12).

Figure 3.11 We need to try to replicate the type of activity a 'natural' puppy would be exposed to. This can be achieved in an outside environment.

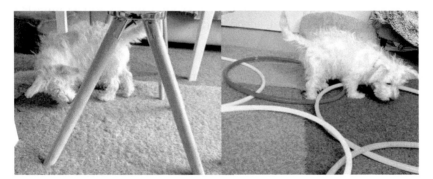

Figure 3.12 We can replicate natural actions and activities inside, even in the smallest space using additional equipment or normal furnishings.

A secure and balanced puppy requires:

- A good **socialisation** programme for brain and emotional development.
- A good physical development for their body.

This enables it to grow into a secure and balanced adult.

If a puppy has the wrong type of physical development, it can lead to an unbalanced body or an asymmetric (**not** symmetrical) body that can have ongoing painful effects within their whole structure.

A balanced body in physical terms means that the left side of the dog has good muscle tone, and the right has equally good muscle tone. It also means that the hindquarters have the same muscle tone as the forequarters.

If the muscles are balanced, they will create equal forces through the muscle and fascial systems, asserting a balanced load over their entire skeleton.

ASYMMETRICAL MUSCLES = ASYMMETRICAL SKELETON

Like all robust structures, it starts with good foundations. The skeleton is the body's foundation, but it is the muscles that hold the bones in position and maintain the integrity.

The muscles and fascia together with the tendons attach the muscles to the different bones, hold the body together, and also convert the joints into moving levers to create locomotion. **And the 'levers' are just that, they 'lever' the body, so their connections must be strong but flexible**.

For a body to be able to create the type of movement that (1) is capable of receiving physical loads and (2) delivering movement drive, including rotation and torque, the joints must be held securely and yet still allow the required flexibility, within the anatomical span of the joints (Figure 3.13).

If there is good stability surrounding the joints, then the dog can safely propel themselves, knowing that the **foundation muscles** will maintain the joint's integrity, strength, and absorb concussion, whilst allowing the **power muscles** to provide a full range of movement, speed, and flexibility.

Take offs!
The foundation muscles are for safe and secure take offs, allowing **power muscles** to power! (Figure 3.14).

Landings!
We cannot see or generally feel the deep foundation muscles as well as we can see the superficial muscles (Figure 3.15) (see page 59). The superficial or power muscles

Figure 3.13 The body needs good foundations, just like every other structure, because if they do not, they are insecure and will eventually collapse. A good example is that of a house, we need to put in excellent and secure foundations so that the building can withstand wear and tear, movement, and maintain its safe integrity for many years.

tend to be the large, long muscles, allowing for large amounts of contraction. Set between two joints that are far from each other, they will facilitate a long range of movement. These power muscles are the ones that change shape and bulge during movement due to the contraction of the fibres (see fibre contraction, page 54).

See Figure 3.14.

Figure 3.14 Take offs! For a dog to drive and power forwards, the body needs the stability and protection of balanced foundation muscles, maintaining the joints integrity in all directions. A dog is shown to be powering off and turning right, whilst holding their body weight as well as powering off from their pelvic region.

Figure 3.15 The foundation muscles are critical for safe and secure landings, facilitating absorption of the concussion, then through the maintained stability of the joints, allowing the power muscles to power! A picture of a dog landing, demonstrating how the load has to be absorbed through their forelimbs, shoulders, and neck.

MUSCLE ACTIVATION

A puppy is born with all their muscles present, but it depends on the type of stimulation they receive from the type of activity they participate in as to whether these **foundation muscles** are recruited by the body to work. Simply, if the puppy does not perform an action that creates a neural stimulus to the muscle, the muscle or muscles will not become activated and they will not fulfil their role.

What activity is required to activate the deep foundation muscles?
Stimulating the deep, foundation muscles requires **slow, multidirectional** movement.

What activity is required to activate the superficial power muscles?
Stimulating the superficial, power muscles requires **faster muscle movement**.
    Without good foundation muscle support and activation, the superficial power muscles dominate the movement patterns and will attempt to perform a double role – both stabilising the joints and providing propulsion, **which will lead to postural dysfunction**.
    When the foundation muscles have not been activated, this power muscle domination becomes an adaptive and dysfunctional type of movement pattern, leading to instability and imbalance within the whole of the dog's body.
    Foundation muscles are for the foundation, and power muscles are for the power – if they work according to their anatomical roles, they will create a body with strong but flexible foundations that will be capable of producing amazing power.
    Why is moving slowly important?
    If a dog cannot perform a movement slowly, how can it then perform the same movement quickly (and safely)?
    The body has to develop muscle patterns through slow movement.
    Moving S-L-O-W-L-Y activates the neural pathways leading to these important foundation muscles (Figures 3.16 and 3.17).
    Slow movement is a way of **targeting individual muscles** or muscle groups. Targeting muscles is essential for activation or neural stimulation.
    Slow movement, by definition, requires varying degrees of muscle engagement from the whole body. To move slowly, the body has to transition through extended phases of suspension. It has to momentarily securely, and often statically, hold a position so that another part of the body can safely leave the ground and transition into the next stride.
    For example, when the right hindlimb lifts to move, the left hindlimb must hold the planted leg securely during that suspension and movement phase. This functional movement demands the standing leg to maintain good joint integrity throughout the whole phase of loading and weight shift in the standing leg. This slow, muscle isolating movement triggers, or stimulates, these stabilising, foundation muscles to become activated and therefore perform their role (Figures 3.18–3.20; driving the potential paths through the neural 'field', page 74).

87

Figure 3.16 Movement patterns, or kinetic chains, if practised slowly, through specific activities, such as the 'enriched environment', will prepare the dog to use movement functionally. The pictures show how forelimb movement patterns or kinetic chains that are developed slowly can then be used at speed.

Why are multidirectional movements important?

Multidirectional movements are critical to building your joined-up puppy. Within our current lifestyle and living and working environment, when we take our dogs out for a walk, we put them on a leash and customary and traditional training suggest that they walk beside us. When, or if, we let them off the leash, they may run or jog away to go and investigate the surroundings.

All these actions are predominantly in one plane of movement (sagittal plane) forwards. This activity and movement recruit the muscles and fascial lines that propel the dogs forwards in a straight line. This type of activity predominantly stimulates or activates the power muscles of the puppy due to the speed of walking, and bypasses activating the foundation muscles (Figure 3.21).

However, if a dog needs to turn, they need their 'sideways' muscles to work to keep stability when they are cornering or turning.

These cornering or sideways muscles form a huge percentage of the joint and limb stability within a puppy and mature dog (Figures 3.22 and 3.23).

Think of these 'sideways' muscles like stabilisers on a bicycle. It gives the bike stability when turning, as the stabilisers keep the bike upright. The bike generally goes forwards but when it turns it needs to stay upright but the bike continues a 'true' path throughout the corner or bend.

Figure 3.17 Movement patterns, or kinetic chains, if practised slowly, through specific activities, such as the 'enriched environment', will prepare the dog to use movement functionally. The pictures show how hindlimb movement patterns or kinetic chains that are developed slowly can then be used at speed.

This is obviously not a very close analogy, but it is intended to demonstrate the importance of 'sideways' stability.

When the dog is moving forwards in a straight-line trajectory, the 'sideways' muscles are not required, therefore not activated.

These 'sideways' movements are called:

- Legs away from the body = abduction.
- Legs towards the body = adduction.

These movements are vitally important to include in the puppy development programme, as these 'sideways' muscles are major contributors to the total stability of the puppy and mature dog.

Figure 3.18 The picture shows the left hind and the right forelimb supporting the whole body through a suspension phase then through the weight shift of the left forelimb and right hind limb taking a walking stride forwards. The head is also fully extended, creating even more load through the forelimbs, demanding even more strength and stability from the foundation muscles.

Figure 3.19 Puppies playing naturally and performing multidirectional actions that encourage balanced muscle development. This picture demonstrates the hindlimb lateral movement.

Figure 3.20 Natural puppy play using their multidirectional movement ability through the forelimbs. This picture shows forelimb medial movement, taking their limb towards the body.

Figure 3.21 A dog running forwards on the sagittal plane is the more repeated form of mobility, but this repeated activity does not form balance within the myofascial system as it predominantly uses and recruits the power muscles.

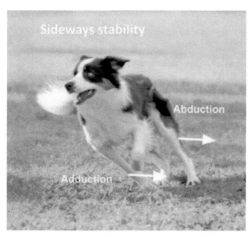

Figure 3.22 The picture shows a dog cornering. To maintain stability and strength to perform this action securely requires the recruitment of the foundation muscles that abduct the limbs (limb moving away from the body) and adduct the limbs (limb moving towards the body). These lateral and medial movements need to be stimulated early in a puppy's life.

Figure 3.23 This bike has stabilisers that help prevent the bike from falling over sideways when cornering in a similar way that good foundation muscle development does in a puppy and mature dog.

The foundation muscles that perform these abduction and adduction actions create a combined support mechanism with the foundation muscles that perform protraction (the limb being taken forwards) and retraction (the limb being taken backwards). The body then has full support through every angle of movement (Figures 3.24 and 3.25).

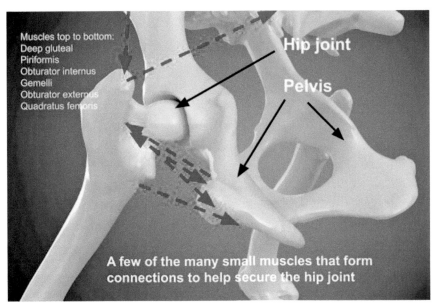

Muscles top to bottom:
Deep gluteal
Piriformis
Obturator internus
Gemelli
Obturator externus
Quadratus femoris

Hip joint

Pelvis

A few of the many small muscles that form connections to help secure the hip joint

Figure 3.24 An anatomical view of where some of the deep foundation muscles are located around the hip joint. The arrows on the muscles are pointing in the direction of their contraction.

Lateral and medial movements are essential within everyday activity.

Figure 3.25 A demonstration of how all these movement patterns are vital for everyday actions that require stability and fluid movement for the whole body.

Activating these deep foundation muscles will not only keep the limbs stable and secure for all actions but keeping the limbs securely supported within the axial skeleton (see page 12) will help to ensure that the vertebrae are not being subjected to unnatural loads and stresses.

These foundation muscles also surround each of the vertebrae, maintaining their integrity and position in relation to each other.

Figure 3.26 Some of the earliest movements a puppy makes are lateral movements, moving sideways to get to the 'milk bar'! When the puppy then leaves the litter, these movements are then often not reinforced and stimulated.

To return to the car analogy in the previous chapter, the foundation muscles ensure the dog's 'chassis' is secure and that the wheels are all tightly attached to the main body.

Puppies, when they are born and suckling, use multidirectional movements to adjust their position. In most puppies, these foundation muscles have already been stimulated, especially if they have been bred by a good breeder who introduces different types of activity, to promote good muscle development once they are up and moving around (Figure 3.26).

When puppies are interacting and playing with their siblings, who are the same age, size, and strength, they will be naturally developing these foundation muscles.

It is when we take them home and put them within the human lifestyle and environment that these movement patterns generally change. They have larger environmental obstacles to cope with, often a larger stronger dog to live with, and the concept of going out for a walk in one direction. They may have had the preliminary development of their foundation, but it is vital it is continued, or their progressive development could become suspended or discontinued.

The puppies require these stimuli to continue regularly and habitually up to when they are at least six months old, but these should also be included within their activities to maintain their physical health for the rest of their lives.

DEVELOPING A FUNCTIONAL KINETIC CHAIN

The kinetic chain fundamentally coordinates and integrates all the structures that connect to form locomotion. It consists of fascia that connect with the muscles (both power and foundation) and to the joints and skeleton to enable coordinated movement.

The kinetic chain also ensures that the body is aware of each moving part; for example, the forelimbs and the hindlimbs are synchronised, creating a body moving as one connected network.

The main conduit for this chain is the fascia. Fascia is the connection that is common to all moving parts; it binds the body to form supported tension together with symbiosis. Fascia is a major ingredient in forming the 'joined-up puppy'.

'Fascia could be thought of as being akin to a body's internal "wrapping" in that it physically connects all the parts of our body, to provide support, communication and physical coordination' (see fascia, pages 53 and 60) (Figure 3.27).

Fascia interweaves with the whole body, it surrounds muscles, joints, and organs, and is seamless.

The interweaving around and through muscles is achieved by its different thicknesses. The thicker, more robust layers of fascia are organised into fascial

Figure 3.27 Young puppies, still in their litter group, playing with everyday objects, giving them the opportunity to discover and explore both with their minds and bodies.

chains (physical connections) that connect with muscles and muscle groups to form kinetic chains (movement connections) to achieve coordinated body movement.

An analogy to explaining fascia could be how, when wearing a shirt and you lift your arm, the shirt will also lift, allowing your arm to fulfil its full range of movement. If, however, you hold the bottom of the shirt down, and try to lift your arm, the held shirt will resist the movement, so the action of your arm will also be restricted. In this scenario, the restriction has nothing to do with muscle function, only demonstrating how fascia facilitates and can equally restrict movement. This fundamentally is how important and expansive fascial connections are, and therefore, how influential fascia is for optimum movement within the whole body (Figure 3.28).

Healthy fascia throughout the body should be in a state of 'appropriate tension' (the bridge, see page 62). Appropriate tension is achieved by the body maintaining a good posture through having good symmetry and balance.

When a body has a well-balanced posture, the kinetic chains will work efficiently, giving the body optimum strength.

If, however, the body is not balanced and does not have a good posture, the kinetic chain will be compromised, and lack strength.

This overall strength does not come from one muscle or even a muscle group, it comes from the combination of linking the foundation and power muscles through the fascial chain that creates an effective kinetic chain (Figures 3.29 and 3.30).

Figure 3.28 Fascial or myofascial movement, through the functional lines or connections, in one part of the body affects the whole – just the same as if you lift your arms, your clothing moves with you.

Figure 3.29 Natural movement patterns encourage strong and stable functional lines to develop.

Figure 3.30 Slow, multidirectional movement with their head down encourages them to 'use' their whole body through the functional muscle and fascial connection, or myofascial connections.

To maintain the integrity of the fascia, the puppy's activities should not include prolonged unnatural postures or repetitive activities, as both of these will have disruptive effects through the kinetic chain.

If there is too much physical load over one or more areas, the fascia's smooth flow of movement over the body will be compromised, or even disabled.

With fascial dysfunction, the whole functional chain (both the muscles and the fascia) will adapt to try to reorganise body load and tension. This often is achieved through physically disengaging kinetic chains, creating weakness over regions of the body (Figure 3.31).

In conclusion, to maintain the correct muscle and fascial tension encompassing the whole body, we must encourage a puppy to use their natural body posture and

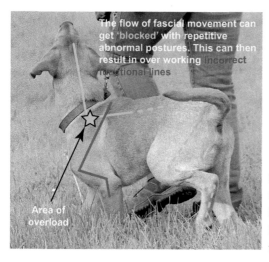

The flow of fascial movement can get 'blocked' with repetitive abnormal postures. This can then result in over working incorrect functional lines

Area of overload

Figure 3.31 if there is repetitive unnatural movement or posture, myofascial lines can be compromised, encouraging a change in muscle activation through restricting functional lines.

natural movement patterns. This will assist good muscle activation to allow muscle development and create healthy kinetic chains and muscle patterning critical for a well-supported, 'joined-up' puppy, which will also have robust foundations to support a healthy mature dog.

FAT

Professor Alex German states that 'overweight dogs are a welfare concern… one third of puppies to adolescents are overweight'.

Overfeeding a puppy and allowing them to become overweight is detrimental to their health and mobility.

Within this chapter, different tissue types have been explored within the context of their healthy development for our 'joined-up puppy'. However, there is one tissue type that must also be considered, but from the context of NOT allowing its development, and that is adipose tissue, or fat tissue.

'Don't worry, it is only puppy fat' is one of the most incorrect and potentially damaging statements spoken about the body condition of a puppy.

If you allow your puppy to gain too much weight, you facilitate the production of fat cells within the puppy's body. These cells will remain in their body through to maturity and the rest of their life. This fundamentally means that these fat cells or adipose tissue will give your dog a lifelong propensity to being overweight.

The body is just like any other engineered structure; any structure is designed to carry the load specific to its design. If that load is continuously exceeded, it will collapse (Figures 3.32 and 3.33).

A puppy carrying too much load or weight for its structure, especially when that structure is in its formative stage, will compromise its development, which is particularly worrying as it is intended to last them a lifetime!

Figure 3.32 Another bridge analogy. If a bridge does not have the correct construction, it will eventually buckle and collapse under the load.

BEFORE   AFTER

Figure 3.33 Before and after pictures of a dog, demonstrating the amount of weight lost through the comparison of 1 kg bags of sugar.

POTENTIAL ISSUES WITH AN OVERWEIGHT PUPPY

- They will have greater issues gaining a good muscle activation and kinetic chain, therefore reducing their potential to develop good body symmetry and balance.
- They will be overloading developing bones and joints, potentially creating abnormalities.
- They will have a high likelihood to develop into an overweight adult. Studies have shown that they could potentially have **two years reduced from their lifespan**.
- They will have a 'bad relationship' with food, having excessive hunger potential, due to hormone imbalances created by excessive fat tissue.
- They suffer secondary conditions, such as earlier onset osteoarthritis.

(a)

(b)

Figure 3.34 A **fat puppy** will often become a **fat adult** that will have many negative health impacts. A brachycephalic, such as a pug, is now one of the most likely to become overweight as a puppy.

It has always been said that the Labrador was the classic 'fat' breed, but studies have been completed by Liverpool University stating that many of the brachycephalic (short-nosed, domed head) breeds are now equal if not higher in a tendency for obesity (Figure 3.34).

The dog in Figure 3.33 lost 12 kg of body fat. To give perspective, she was carrying the equivalent of 12 kg of sugar. Conduct your own experiment, carry just 1 kg of sugar or rice around for a day, feel that load of 1 kg additional weight on your body, then consider the impacts of your puppy or dog carrying that or more weight in the form of fat over a longer period.

Remember, every kilogram extra = extra work for the dog's whole structure to carry; for the heart to pump blood and for the whole body to maintain.

WHAT CAN WE DO TO PREVENT OUR PUPPIES FROM BECOMING OVERWEIGHT?

Professor Alexander German's advice:

1. Do not overfeed.
2. Be aware that feeding 'treats' for treating or training should come out of their daily allowance.
3. Weigh your dog regularly.
4. Encourage scavenging exercises so that they hunt for food.
5. Use a puppy development chart to plot healthy weight gain.

SO MUCH MORE TO FAT THAN WE REALISED!

For many years, fat was thought to be just an unwelcome, unhealthy tissue that was detrimental to our body through overburdening the heart and vascular system and overloading the skeleton and joints. That is still true, but there are many interesting studies that are discovering that fat tissue can actually produce hormones, which can give the brain corrupting signs that influence eating habits.

In conclusion, allowing your puppy to become overweight will leave them vulnerable to the potential of a huge number of health risks including secondary health conditions such as osteoarthritis.

CORE STRENGTH AND PROPRIOCEPTION: WHERE DO THESE FIT WITHIN DEVELOPING A PUPPY?

Core development or core strength and proprioception are words that may be seen within various fitness programmes for dogs (and people). It can seem all very confusing wondering how all these aspects of building a body fit together.

There are fitness programmes that state you must 'strengthen your dog's core' or 'how to improve their proprioception'. The fact is that these **cannot be done effectively in isolation**, it must involve the body in its totality.

We should be moving away from just 'core strength' and we should be looking at developing functional strength through functional movement, which is developing natural movement patterns, something that has been discussed throughout this whole book.

If the whole dog's body is balanced with equal and appropriate tension, the dog will have active core muscles and full body awareness (=core strength and proprioception).

WHAT IS MEANT BY THE DOG'S 'CORE MUSCLES'?

The exact location of the 'core muscles' is often open to slight interpretation. Fundamentally, it comprises the epaxial and hypaxial muscle groups (Figure 3.35).

The **epaxial muscles** are situated between each vertebra = multifidus (primarily).

The **hypaxial muscles** are generally agreed to be primarily the abdominals and obliques.

Core strength was considered to be balanced strength and tone between the epaxial and hypaxial muscles, so that when they are working together, they form stability for the axial skeleton (the vertebrae).

However, if you add the fascial function lines, it can then be seen that the fascia joins up these two components of the core muscles, and also the foundation and power muscles. Together, this forms a more balanced and stronger dog because the muscular system (including the core muscles) has been coupled up with the fascial connections to form a **synergistic biomechanical movement** (see Figure 3.35).

WHAT IS PROPRIOCEPTION?

Proprioception is almost like an enhanced spatial awareness. A well-connected dog, or one with good proprioception, is aware of the location of each of their legs in relation to each other and is able to move each separately (Figure 3.36).

Figure 3.35 Demonstrating where the dog's core muscles would be situated on their body.

They are under control and connected to where their body is in relation to their environment, to the point of knowing how much strength and movement is required to perform each action; it also could be thought of as 'muscle memory' (a puppy that is overly clumsy could be considered as having undeveloped proprioception, or not being 'joined-up').

This connection is a complexity of the body combining together, which will be through the nerve connections and also the myofascial (muscles and fascia) network.

NB: It is known that the fascial system supports its own nerve supply.

Myofascial development and proprioception work hand-in-hand with our joined-up puppy.

We are developing the whole body from a good foundation, ensuring the legs (appendicular skeleton) are securely attached to the axial skeleton. This enables the axial skeleton foundation muscles to activate. Adding the facial supporting tension will provide this well-connected body with the stability required for both static and dynamic, 'spring loaded' movement (Figure 3.37).

OLD BELIEFS VS CURRENT BELIEFS

Original thoughts were that the skeleton was the frame that the body was built on and it was through the skeleton that the body was supported.

Whereas current beliefs state that fascia is the actual major mechanism that supports the body through its spring-loaded tension or tensegrity (see page 63).

Figure 3.36 Marge, a nine-week-old puppy, demonstrating good proprioception by being able to negotiate the hoops easily, knowing exactly where all her feet are and will be placed.

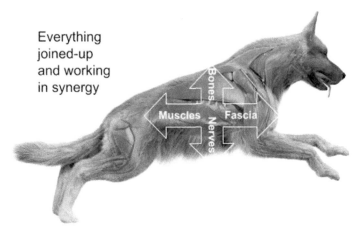

Figure 3.37 For fully joined-up movement, the skeletal, neural, and myofascial (muscle and fascial) systems have to work together

A good analogy we could use to compare these two beliefs are two different types of tent:

- The *ridge pole tent* is held up and secured by a solid frame (Figure 3.38).
- The *pop-up tent* asserts lateral force to maintain its integrity (Figure 3.39).

Figure 3.38 A ridge tent, secured with a solid internal pole structure.

Figure 3.39 A pop-up tent, secured through internal lateral force.

Contemporary views are that our bodies are held together using the spring-loaded concept of tensegrity, very similar to a pop-up tent:

- The ridge pole tent is very strong, but not mobile or flexible (and needs external stabilisers).
- The pop-up tent is very strong but also mobile and flexible.

However, whatever your view on how the body is supported, within each scenario, if one of the corners is not secure, the whole structure will lose its integrity.

Therefore, whatever you believe to be the correct structure and supporting mechanisms, we should be ensuring that our puppies' limbs and limb attachments are fully supported, stable, and secure.

KEY POINTS

- A healthy and strong body comes from the whole, not individual muscles.
- For locomotion, all the relevant systems have to be activated and synchronised to function efficiently.
- Correct posture when moving is vital to allow the body to develop strong functional movements using the correct muscle patterning.
- Correct posture when moving is vital to activate the correct muscle, that will allow the body to develop strong functional movements, using the correct muscle patterning.
- If these natural movement patterns are encouraged in a puppy, they will be established for them to take into maturity (*subject to lack of untreated injury, overuse, or repetitive strain*).

4   Building your puppy using the Galen Puppy Physical Development Programme©

O ne of the first things we are going to have to do is have a radical change in thinking about walking or exercising our dogs and especially our puppies. When you take your dog out for a walk, you should be considering *their* needs over yours. The walk should be biased towards your dog or puppy and what they need to fulfil them physically, mentally, and emotionally (Figures 4.1 and 4.2).

We have looked at the importance of the constituents that form the joined-up puppy, brain/nerves, bone, muscle, and fascia. We now need to consider the activities we can ask our puppies to participate in that will promote these tissues to 'join up' or to build our puppy into a robust but flexible adult.

Dogs in the wild or undomesticated 'street' style living do not habitually, if at all, walk one hour nonstop a day. They are more likely to take **short bursts** of varied generally low impact activity, usually within close proximity; activities such as playing, pseudo or real hunting, exploring, scavenging, and, of course, sleeping.

The concept of taking long and extended walks has become part of our human culture when living with dogs. The single long walk can also include rapid, high impact, and highly concussive and overly stimulatory repetitive strain promoting activities such as lead pulling and ball, frisbee, or stick throwing. These activities are either encouraged or not discouraged, no doubt driven by good intentions. However, these repetitive and excessive exercise routines can be **highly detrimental** for our dogs (see lead walking and ball throwing on page 189).

These excessive and potentially high impact activities during one or two parts of the day can then be followed by limited, or little activity or stimulation, for the rest of the day. This is contrary to what a puppy (and an adult dog) needs to stimulate their physical frame and meet their emotional requirements. See page 6 for a **comparison between how dogs lived alongside humans and how they live now within a human environment**.

Historically, we have been indoctrinated with 'exercising' or 'walking the dog'. It has almost become task driven; and almost something of a prerequisite of a human's requirements from a dog they have living with them. Dogs should not be considered just a walking companion, they are a 'whole life companion'.

Dogs and puppies love being involved with every part of their human world; wanting to have the opportunity to investigate situations or objects; be given the opportunity to explore safely and develop mutual boundaries based on their environment, as they would naturally; and not be shut off or restricted from so much of their potential family life.

107

DOI: 10.1201/9781003268789-4

Figure 4.1 A human loves to stay in touch through their mobile phone, a dog likes to stay in touch with their nose… and maybe our dog's perception of a walk is very different to ours? Marshall's Law by Tomoyo and Roger Pitcher, used by permission, all rights reserved.

Offering the correct physical exercise, activity, and inclusion can reduce so much frustration and fear that could otherwise lead to all sorts of undesirable behaviours exhibited by puppies and dogs.

THE PUPPY ACTIVITY PLAN

We start this section by exchanging the word 'exercise' for 'activity'. Dogs and puppies need activities that are appropriate to them. Just like us, they need activities that will stimulate their bodies and brains in a fulfilling way, which will develop their minds and bodies and give them purpose.

The word **exercise** is synonymous with tiring your dog through taking your dog for a walk, but not necessarily connecting the importance of these outings with your puppy's, or dog's, vital physical and psychological development, stimulation, fulfilment, and inclusion.

Putting together a good 'activity' plan for puppies is not easy, as everyone lives in different environments and with differing situations. However, if we could adapt a mechanism that can be used for most situations, it would make it much easier to understand and apply.

Common advice for a puppy's guardian is 'five minutes "exercise" per month of age (up to twice a day) until the puppy is fully grown'.

This is practical, easy to follow advice, but it can be and is so often *wrongly interpreted*. The mechanism of progressing time and activity with age is an easy one to follow.

However, if we just adapt it a little, then it could formulate an amazing, more rounded, development programme for your puppy.

This Galen Puppy Physical Development Programme includes different activities to stimulate areas of a puppy's physicality to help activate and therefore develop their body **as a whole**.

It also subdivides activities into the **type of impact** different activities will have on the developing (and also mature) canine body. Categorising 'impact' from productive to destructive

For the body to develop, it needs to be challenged, but these challenges should be appropriate to the canine's functional anatomy. It will **then** provide positive skeletal impact through the correct planes of movement, providing authentic and natural healthy loading through joints, bones, and soft tissues connections (muscle, fascia, tendons, and ligaments).

Figure 4.2 A person going on a walk at their speed with their perceptions of a walk, without necessarily considering their dog's perception of the same walk.

WALKING YOUR PUPPY

Before we discuss and look at activities that will optimise your puppy's development, we need to look at the basics, such as 'what is walk?'

Something as seemingly basic as 'walk' can be performed in a way that is incredibly beneficial for your puppy. Conversely, common mistakes can make a 'walk' highly detrimental!

The association with walking and having dogs is inextricable. However, within a walk we have the potential of creating excessive and damaging physical and emotional stresses on our dogs. Whereas if we changed just a very few things, we could offer them a fulfilling, positive, bonding, and healthy activity that would maintain their health and condition, with the potential to give them extended longevity.

Part of what we need to do is think like our dog or puppy. We must consider their size or scale in relation to us. Also, consider **their** perception of the totally alien environment you are introducing them to, and your idea of **a walk** could be miles apart!

Often it is thought that a quick walk 'around the block' is so they can urinate and defecate and 'stretch their legs', but if you asked a dog, maybe they would say, the best bit is catching up on all the local gossip and events (Figure 4.3).

We can't see all this information our dogs are connecting with because we cannot see or indeed smell the scent particles that your dog can. And they need a good sniff, not just a passing waft! According to the University of Lincoln's Professor Daniel

Figure 4.3 Young puppies should be encouraged to explore and use their innate scenting abilities.

Mills, a dog needs at least **ten seconds of concentrated sniffing** to gather all the information they want and need. If this is interrupted, they have to start again. If we are being considerate to our puppy's needs, we need to allow them ten seconds to absorb and evaluate the scent (Figure 4.4).

It would be a bit like us going to see a great film that was so exciting, then being dragged out before the end… and worse still, no one else has seen that film, so you will never know the ending! (Figure 4.5).

Figure 4.4 Scenting skills develop as the puppy grows, allowing them time to understand and explore their environment.

Figure 4.5 People watching a film would not want to leave watching something exciting before the end… so why remove your dog from their 'film' of sniffing?

THEIR LEGS ARE SHORTER THAN OURS!

We talk about taking our dogs for a walk, but whilst we may be walking our dogs are actually trotting. This is primarily because their legs are shorter than ours, so they have to trot to keep up! (Figure 4.6).

There is a huge difference in our range of movement (or stride length) compared to our puppies (and also mature dogs). Therefore, we are conditioning them to trot; or worse still *removing their innate ability* to actually walk.

In a natural environment a dog would walk so much of the time. They would use their walk gait for many activities.

Walking is important for so many reasons:

1. the walk is a four-time gait – what this means is that each leg moves separately and individually. This means that each time the leg hits the ground it must have the muscular capability of absorbing the force and propelling it into its next stride (Figure 4.7).

This is important because:

- It helps develop the foundation muscles individually within each limb and joint.
- It embeds good proprioception and spatial awareness through developing the connection between brain/nerves, bone, muscle, and fascia.

Figure 4.6 A puppy's legs are shorter, therefore, their stride length is always going to be shorter than ours.

Figure 4.7 A four-time walk when each leg takes the full load during one part of the transition.

- It is a good natural rehabilitation exercise for gently but globally exercising and conditioning the whole body.
- An active walk is a good conditioning and fitness gait (for mature dogs).
- It can also help to identify an uncomfortable or dysfunctional region, e.g., finding a lameness.
- Cross-lateral movement is important for the brain to body connection.
- For mature dogs, it is highly effective for warming-up and warming-down, pre- and post-event.

The trot works across the diagonal of the puppy, so diagonal legs share the load and absorption, which makes it easier and faster to perform to keep up with their human (Figure 4.8). Also, if the puppy or dog has any discomfort in any region (Figure 4.9), the trot enables less focus or load on the uncomfortable area or region; **this can also disguise an issue**.

This gait is the one that most puppies use to keep up with our stride length; this is too fast for puppies to develop good muscle structure and development.

Figure 4.8 A trot, which is a two-time, more economical gait.

Figure 4.9 A pace, when a dog looks like they are walking when their right legs and left legs move together. A 'pace' can often indicate an avoidance of a discomfort or a dysfunctional biomechanical chain.

FIRST THINGS FIRST

When we first walk our puppy, we must be very aware of so many things. We must remember how huge the outside world, and everything in it, must look to our puppy, and how noisy and strange it must all feel, including underfoot.

Even if your socialisation has been extensive, puppies are still going to find so much of it alarming, from moving vehicles, noises, people, children, and of course other dogs. In these scary situations, they are going to inadvertently pull into their restraining device (the restraining device being a collar and leash or a harness and leash).

Consider how it will feel to your new puppy if suddenly, they encounter something that concerns them and then they immediately experience pain from the jolt of a collar around their throat, or a constriction around their body with an incorrectly fitting harness.

These first impressions will stay with them, and quite likely, you will then have to manage them as an ongoing issue. It would be better that you ensure their experiences are as good as they can be. (We all will make mistakes that we have to later correct or manage, this is more about trying to alert the reader to as many things to avoid as possible.)

Ensure you use the correct equipment for your puppy, and that it fits around their anatomy so they can move and use their body without impingement, future pain, or musculoskeletal damage (collar vs harness, see page 207).

HOW FAR TO WALK YOUR PUPPY

One of the most common and generally unintentional mistakes we make with our puppies is walking them too much (see page 120 of the Galen plan and page 187 on overwalking your puppy). Overwalking or over exercising your puppy can stress them both physically and mentally, their bodies are undeveloped and they need careful moulding. Their bones are still growing and are partially cartilaginous (growth plates, see page 78). The joints and long bones need concussive forces to help stimulate and develop them, but limited, overly stimulation can do harm to these developing regions.

We also want our puppies to develop good muscle and fascial connections (see page 80). These need to develop primarily through slow multidirectional, multiplane, and multispeed movements, which will help them to develop good soft tissue connections and provide deep foundation muscles to help stabilise the joints and skeletons. If they are overwalked, these processes could develop the wrong muscle patterning, which does not create a puppy that is strong and stable for the inside.

Over exercise also will cause them to be tired, no one would expect a toddler to walk for more than five or ten minutes! They may not behave as if they are tired, but they are basically stressed, both physically and mentally and find it difficult to relax, so they become a turbo puppy!

POST-WALK TURBO PUPPY!

POST-ACTIVITY CALMING METHODS... PLEASE READ!

There is often a misinterpretation with regard to exercise and your puppy. When you take your puppy or adolescent out for a walk, they may meet another dog and run excitedly, maybe further and longer than normal, or recommended; or participate in an ultra-high-impact activity such as chasing after a ball. After such events, when the puppy gets home, they can appear to be still full of fun and energy; this can lead to the belief that the puppy needs more exercise to 'tire them out', but so often this is not true (Figure 4.10).

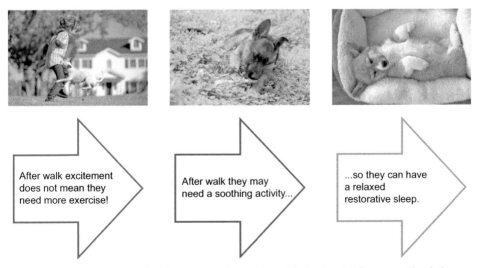

After walk excitement does not mean they need more exercise!

After walk they may need a soothing activity...

...so they can have a relaxed restorative sleep.

Figure 4.10 The importance of understanding that exuberant behaviour is often over stimulation.

They do not need more exercise. They need calming – a settling activity or activities to allow them to calm down. Puppies find it difficult to self-regulate, so we have to help them with coping mechanisms to calm themselves. Because what you are actually seeing is an **adrenaline-driven puppy**.

Adrenaline is a powerful hormone and can convey a picture of a puppy bursting with natural energy, but your puppy could be really tired. Further exercise on a tired, unfocused puppy could be highly detrimental, both physically and psychologically.

The use of **settling activities** is a valuable tool for puppies and young and mature dogs. It calms them and can prevent them from continuing activities or exercise that could over-stress and potentially damage them (see calming activities, page 151).

This is such an important aspect of rearing a puppy. Recognising when they need to be calmed and not further excited. If this can be achieved, your puppy will be more content and so much easier to manage and can live in greater harmony.

Some of these de-stressing activities are also part of the Galen Puppy Physical Development Programme. The puppy can de-stress and develop its structure, rather than continue in potentially destructive physical activity.

KEY POINTS

- Puppy empathy, how may a puppy perceive each situation from their, not your, perspective.
- Allow your puppy to be a canine by encouraging their development of natural instincts and movement patterns.

5 Categories involved with the puppy's development

The foundation of this programme is to develop these three key components within the puppy's activities during the day (Figure 5.1):

- Brain and neural pathways (connect brain to body and body to brain).
- Skeletal development – ensure strong anatomically true bone growth and joint development.
- Muscle, fascia, tendon, and ligament development to build the integrity and strength of the body as a whole.

All three categories need to be equally developed for the puppy to be fully physically 'joined-up'. Therefore, the time given to these during the day has also got to be relative to the **type of exercise** from a physical and psychological perspective. Stimulate but avoid overstimulation.

Skeletal development is important as this requires appropriate impact through activity and exercise. Too much can be detrimental, but too little can also be detrimental.

The type of activity needs to be balanced against muscle, fascial, and neurological development. The skeletal system development is often ahead of muscular and connective fascial development. Hence, the classic long-legged puppy, with the disproportionately large paws! That puppy is growing in height and stature but its myofascial connections are in a 'catch-up' state and will not be able to fully support the skeleton structure.

Therefore, physiologically, it is critical to construct a body development plan that is sympathetic to the different rates of growth between all the body structures. Offering the correct activities is required for these rapidly developing bones and joints alongside the correct activities required to develop the **underpinning support structure of soft tissue**.

As has been discussed (page 71), we need to develop the puppy's physical balance. However, within our current environments, so much does not facilitate *natural development*, and has a highly negative impact on our dogs and puppies (see page 6).

Therefore, it is our responsibility to ensure that we offer and encourage our puppies to develop each of their physical components to ensure that their bodies can function at their optimum which will then help to:

- **Reduce the risk of accidents** because the puppy and dog are aware and **have control of all four limbs**.

DOI: 10.1201/9781003268789-5

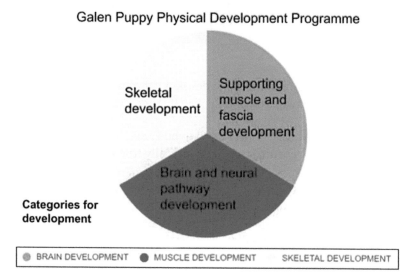

Figure 5.1 Looking at the categories that are involved with the puppy's development.

- Ensure optimum joint stability, **reducing joint wear** and muscular compensation.
- Maintain positive **myofascial strength and connection**.
- Aid optimum biomechanics that **enhance strength and stamina**.
- Provide the optimum framework for a dog to lead a full, **happy, healthy, active, and comfortable life**.
- Use their functional anatomy to maintain optimum balance within functional movement.
- Give them a **longer health span** – healthy longevity.
- Reduce the risk of pain.

It is important to note at this juncture, a healthy fit mobile body is not just about keeping a dog moving, it also facilitates and is a vital component to their total health.

- **Less stress**. Both physical and psychological – stress is a known vector for disease and dysfunction.
- Reduced back and joint problems.
- Free active natural movement is imperative for healthy bodily functions, e.g., digestive, circulatory, lymphatic, respiratory.
- Less pain.

HOW TO USE THE PROGRAMME

Flexible timings – The programme is intended to be flexible so it will be easy and importantly work for both you and your puppy and accommodate whatever environmental space you have.

Indoor or outdoor – Many of the activities can be adapted for either indoor or outdoor use.

Can include your other dogs – Likewise, you can easily integrate or adapt them into other everyday activities, including other dogs in the household too.

PHYSICAL DEVELOPMENTAL ACTIVITIES – DIVIDED INTO *THREE IMPACT LEVELS*

The aim is to achieve the correct level of impact for appropriate skeletal cellular development, together with the relevant loading and directional challenge for soft tissue and connective tissue development (Figure 5.2 and Table 5.1).

Balancing the activity between low and high positive impacts will provide a great foundation for your puppy and mature dog.

It is strongly advised **not to partake in ultra-high detrimental impact** (see pages 117, 118, 119, 121, 124, 129 and 157). If any of these activities are repeated the risk of damage and inappropriate development is high.

		Positive anatomical development		
Impact level	TIME per DAY	BRAIN/ (+NEURAL CONNECTIONS)	BONE + JOINTs	MUSCLE (FOUNDATION)
LOW	up to 10 mins or twice a day - or when they choose to stop	✓	✓	✓
HIGH	5 minutes per month up to 6 months of age		✓	
ULTRA HIGH	Try to avoid completely - especially repetitive application as they can be destructive to good development	✕	✕	✕

Figure 5.2 An example of a good puppy programme.

Table 5.1 A Puppy and Adolescent Programme by Age

Age of puppy	High impact	Low impact/calming activity examples (see chart pages for age appropriateness)	Calming activities – these can be interspersed with low impact activities	Sleep (see sleep is important, page 151)
Two months	**Get expert help with lead walking** Ten mins walking, including free running, twice a day (if they can only go out once a day, then add another low impact activity – **do NOT double up walk time, as that could then become ultra-high impact)**	Ten mins twice a day (examples) Enriched environment Post-meal food search over very low messy poles	Paper shredding Chewing or 'licky' mat Snuffle mat	Require at least 15–18 hours per day
Three months	15 mins walking, including free running, twice a day (if they can only go out once a day, then add another low impact activity – **do NOT double up walk time, as that could then become ultra-high impact)**	15 mins twice a day (examples) Enriched environment Post-meal food search over very low messy poles Vertical poles food search Horizontal poles food search Light tuggy	Chewing a bone – also great for their feet and toe strengthening (page 151) Shredding Finding treats in loo rolls Brain games	Require at least 15–18 hours per day

(Continued)

Table 5.1 (Continued)   A Puppy and Adolescent Programme by Age

Age of puppy	High impact	Low impact/calming activity examples (see chart pages for age appropriateness)	Calming activities – these can be interspersed with low impact activities	Sleep (see sleep is important, page 151)
Four months	20 mins walking, including free running, twice a day (if they can only go out once a day, then add another low impact activity – **do NOT double up walk time, as that could then become ultra-high impact**)	Post-meal food search over messy poles Food search around vertical objects Free exploring different terrains/surfaces 'Find it' food or toy search (ensure food is used from ration to avoid the puppy becoming overweight)	Chewing a bone – also great for their feet and toe strengthening (page 151) Shredding Finding treats in loo rolls Brain games	Require at least 15–18 hours per day
Five months	25 mins walking, including free running, twice a day (if they can only go out once a day, then add another low impact activity – **do NOT double up walk time, as that could then become ultra-high impact**)	Messy poles with food search Enriched environment balancing 'Downward dog' The human gym 'Find it' – use a toy	Chewing a bone – also great for their feet and toe strengthening (page 154) Shredding Finding treats in loo rolls Brain games Tuggy	Require at least 15–18 hours per day

121

(Continued)

Table 5.1 (Continued)   A Puppy and Adolescent Programme by Age

Age of puppy	High impact	Low impact/calming activity examples (see chart pages for age appropriateness)	Calming activities – these can be interspersed with low impact activities	Sleep (see sleep is important, page 151)
Six months	25 mins walking, including free running, twice a day (if they can only go out once a day, then add another low impact activity – **do NOT double up walk time, as that could then become ultra-high impact**)	Messy poles with food search Enriched environment Balancing 'Downward dog' The human gym Find it – use a toy	Chewing a bone – also great for their feet and toe strengthening (page xx) Shredding Finding treats in loo rolls Brain games	Require at least 12–15 hours per day

From six months onwards, stick to the timings and activities for a six-month-old.

NB: It is important to maintain this programme into and beyond adolescence. A dog will commonly continue to grow and develop up until they are two years old.

KEY

The main areas of development:
- Muscle/fascia development
- Neural development
- Skeletal development

- **Low positive impact (muscle/fascia/neural/skeletal)** can be repeated during the day but aim for twice as long as the high impact exercise (Table 5.2).

Table 5.2 Low Impact Enrichment

Enriched environment – Low impact (Figures 5.3 and 5.4)
Must always be conducted on a non-slip surface
Never leave your puppy unattended

Figure 5.3 Enriched environments can be big or small and have a mixture of interesting but safe articles to investigate.

Figure 5.4 Enriched environments can also be outside.

(Continued)

Table 5.2 (Continued)   Low Impact Enrichment

Instructions:

Set up the room without the puppy. Then allow the puppy into the area, but only talk to them to give them confidence, not to encourage them to go in any direction. Allow them to investigate in their own time and own speed.

This can also be done outside but ensure that it is a safe place, with no interruptions from other dogs or humans

Age: From eight weeks:

- This can also be progressional as your puppy grows

Frequency: Every day but do slightly alter the articles for additional stimulation.

For how long? Allow your puppy to decide. You may think they have lost interest after a couple of minutes, and they may start wandering around the outskirts of the objects. Allow them a few minutes to walk around and process. It is a powerful confidence builder if your puppy chooses to investigate of their own volition.

Preparing your puppy: Prepare without the puppy in the room.

Post-exercise:

– If your puppy is very **excited**, it may be a good idea to give them something to chew in a quiet corner. Chewing will help calm them.

– if your puppy is **calm** after it has decided it has finished, it should be left alone to process by being given time to rest or sleep without being disturbed.

This will enable better processing and therefore application of what they have experienced.

Equipment:

- Cardboard boxes
- Loo roll cardboard
- Soft toys (without a squeak is less excitatory)
- Low poles (broom handle size)
- Everyday objects, i.e., vacuum cleaner
- Clothing – socks, etc.
- Different safe surfaces
- Clothing hanging up that they can walk under and through
- Scrunched-up newspaper
- Umbrella
- Shoes
- Large cushions* – these are good to simulate rough undulating ground

Do not put your puppy on a human bed to replicate undulating ground because they could find it concerning and not be able to remove themselves by their own volition, or they could jump off and damage themselves.

Aim: To encourage your puppy to walk over, around, through, into, and under all sorts of miscellaneous objects. Walk slowly and multidirectional over different surfaces, uneven, or soft texture under their feet. Encourage nose work, confidence through being allowed to choose their path.

Aids strengthening of the neck through the natural use of sniffing and lifting the head.

Progression *(as your puppy grows in size and confidence, add more challenging items)*:

- A small tray may be propped to cause a small wobble effect.
- Cushions can be added additional experience with uneven and unstable surfaces.
- Hidden treats can be added into scrunched-up paper or loo rolls, to make the environment even more exciting and allow the puppy to find a reward.
- Larger boxes for them to climb onto/into/find things in.

When your puppy has done this a couple of times, and gained a little confidence, you can incorporate another dog from the family, so they can explore together.

This is great to develop dogs of all ages.

Each of these objects can also be used in isolation.

This may seem like a lot of work, putting it out then clearing everything away – but this will keep your puppy so much happier and reduce destructive behaviours derived from frustration/boredom or even fear.

Table 5.2 (Continued) How the Exercise Will Help Build Your Puppy

Brain/neural pathways	Skeleton	Muscle
Enhances olfactory receptors – enhance scenting.	Gentle movement patterns encouraging healthy bone development.	Encourages slow movement.
Due to natural head carriage, assist neural pathways to develop foundation muscles.	Synovial production	Encourages natural multidirectional movement/ cross-lateral movement.
Introduces the puppy to 'choice', enabling their free will to explore and process the environment and objects at their own pace.	Joint usage. Develop good foot and toe development through the use of different surfaces.	Encouraging isolated movement through their pelvic region and hind legs, activating their deep hip stabilisers or foundation muscles.
Socialisation and habitualisation (introducing your puppy to everyday objects).		Engaging kinetic chain through natural head position.
Participating in an activity that is instinctive will therefore allow them to naturally relax and sleep.		Introducing the body to uneven and soft surfaces.
Improve the bonding between you and your puppy due to the positive connections formed from your passive involvement with a fun and fulfilling activity.		Naturally enhancing coordination and proprioception.
		Develops their foot placement skills, enabling them to be more sure-footed, so they are able to successfully negotiate undulations, hills, steps, and terrains with secure foot placement.
		Develop balance and engagement of the tail.

Two months = 20 mins per day (divided through the day).

Three months = 30 mins per day (divided through the day).

Six months + = maximum 60 minutes of activities into adulthood.*

*These low impact activities should be part of your dog's life for their entire life.

Many of the low impact activities cover all the development categories, as well as being a calming activity.

- **High positive impact (skeletal)** should be implemented twice a day at five minutes for each month of age of the puppy
  - Two months = 10 mins per day, twice a day
  - Three months = 15 mins per day, twice a day
  - Four months = 20 mins per day, twice a day
  - Five months = 25 mins per day twice a day
  - Six months - 30 mins per day twice a day

This should be the maximum amount of exercise for a growing dog. This is also sufficient for mature dogs, together with low impact activities.

- **Ultra-high detrimental impact –** *this should be avoided* **or kept to an absolute minimum**.

  Continued, ongoing detrimental activities for extended periods* of time is highly likely to cause body distortion, compaction, and concussive issues.

  *A few minutes every day for over six months, or maybe even over less time could **cause permanent damage** (see page 190).

Sleep – sleep is a vital ingredient for healthy puppy development. For a puppy to grow into a balanced and healthy dog, both physically and mentally, they need to receive enough restorative sleep.

Restorative sleep allows the brain to process and organise all the thoughts and events of the day – to aid them to make sense of all their learning. The body requires sleep to allow positive cellular activity, facilitating healthy growth and development.

If a puppy is not comfortable, whether it be physically in an inappropriate bed or situation, or mentally in a non-conducive environment, and is interrupted during their sleeping, it can result in destructive or aggressive behaviour, just in the same way we can be grumpy when sleep deprived.

'The majority of incidents involving aggression are likely to happen on Sunday afternoons when dogs will be at their most tired after a long weekend' (Bailey 2021).

For puppies that are growing at such a fast rate, it is vitally important they receive the correct amount of sleep in a comfortable bed in a 'safe' environment (see bed, rate of growth, page 178, Table 5.1).

Safe in this context means an emotionally secure environment and does not imply crating.

THE DOS AND DON'TS OF WALKING YOUR PUPPY

Two sides of the same activity; how 'walking' can be a productive and fulfilling part of a puppy's life (page 130) but it also has the potential for being restrictive and limiting to natural functional movement (Table 5.3).

LIMITATIONS AND CAUTION
- Potential for overstimulation.
- Impede movement patterns.
- If using a collar and the puppy is pulling, potentially damaging nerves travelling through the neck.
- Change gait style from walk to trot.
- Create incorrect loading on joints and skeleton due to repetitive asymmetric actions.
- If using a collar or an incorrectly fitting harness with the puppy pulling it is potentially damaging muscles in the neck, throat, and forelimb.
- Head position.
- Asymmetric movement patterns from walking consistently on one side.

Table 5.3   Ultra-High Impact Activities When Walking Your Puppy

This table shows how 'walking' can have the potential for being restrictive and limiting to natural functional movement. Walking on a leash has been classed as high impact because of all the potential 'unnatural' components to the activity. To be safe, puppies must wear a collar or harness and leash, but like all equipment, they can cause discomfort and restrict functional natural movement.

Walking too fast for your puppy or dog. The puppy has to walk at the speed of the human. The puppy's legs are considerably shorter than a human's.

The puppy learns to trot not walk. The puppy has to walk at the speed of the human, and learns to trot, losing the muscle patterning for 'walk'.

Puppy looking up for extended periods. When walking to 'heel' the puppy is inclined to look up for extended periods. This is a very uncomfortable position for the puppy (and mature dog) to maintain, creating 'pinch points' in their neck, shoulders and lower back.

For the formative training this happens, but they must not be expected to continue this action (see puppy with blocked fascial line, page 94).

(Continued)

Table 5.3 (Continued) Ultra-High Impact Activities When Walking Your Puppy

Puppy walking on one side of human. As well as the puppy looking up, if they are encouraged to walk on one side of the human, this could develop over time into a power muscle domination creating a 'crabbing' type gait, or not tracking up (the tracks of their hindlimbs do not follow the tracks of the forelimbs). This is because the dog will be walking for prolonged periods just slightly curved through the body, banana shaped. This can also be relevant walking next to children's strollers.

Puppy not allowed to – and physically unable to – sniff. The puppy is not given the amount of time to discover their environment, yanking on the harness to keep moving and not to 'keep on sniffing'. This can create frustration and a lack of autonomy.

This also can relate to a short leash being used, preventing the puppy from physically being able to reach the ground to sniff, as well as general physical restrictions.

Also, the type of equipment worn by the puppy, especially if used to prevent pulling

Puppy wearing restricting equipment. The equipment they have to wear for a walk to keep them safe is vital, but the same equipment could be restricting, uncomfortable, painful, or even damaging

(Continued)

Table 5.3 (Continued)   Ultra-High Impact Activities When Walking Your Puppy

Puppy pulling persistently on the leash. Be allowed to pull into their collar or even a harness, especially if it is not fitted properly; these can potentially create physical damage or restrictions (see problems with leash walking, page 207).

Over walking. Walk further than they are ready for, creating overloading of their muscles, bones, and joints, incorrect muscle activation or development, becoming tired and uncomfortable

Too much stimulation coupled with exertion. Overstimulated by too much running with bigger and older dogs. To walk your puppy longer than recommended can be detrimental to their development. It does not 'wear them out' so they will sleep, it will overstimulate and stress them, both physically and mentally.

With external stimuli, such as meeting bigger or older dogs on a walk, a puppy's exercise should be moderated and play stopped. Extended play and running in this situation are driven by adrenaline not by physical capability.

Table 5.3 (Continued)   Ultra-High Impact Exercise

The impact of ultra impact exercise on your puppy's development How the exercise could potentially harm your puppy		
Brain/neural pathways	**Skeleton**	**Muscle**
Nothing positive	Nothing positive	Nothing positive

WALKING ON A LEASH

High impact – recommended with a well-fitting harness (see page 215, Table 8.1):

- The human should walk with consideration of the puppy's length of leg, therefore the puppy's stride length.
- The puppy must be allowed to walk at their own pace. The walk is a low impact highly developmental and restorative pace. However, by only giving the puppy the opportunity to trot, it will not offer them the opportunity to develop this important gait and muscle patterning.
- To allow the head to be in its natural and scenting position, the harness should allow a natural movement pattern, which is more comfortable and better for physical development.
- Allowing a puppy to walk in altering directions not just in straight lines. Walking in a gentle curve can help engage good muscle chains.
- The puppy should be allowed at least ten seconds at every sniffing point and given some freedom of movement and choice where to walk and sniff.
- Ensure that the harness is a correct fit and use (see equipment, page 207).

Ensure suitable professional and approved training about how to walk safely (not controlling equipment unless you have received varied expert-approved advice; see equipment, page 207).

Stick to the programme of timing high and low impact exercises to achieve the correct balance.

INSTRUCTIONS

It is really important to receive professional instruction on leash walking – a puppy that pulls will assert incorrect loading through their whole body (see resources and recommendations).

When you are leaving home to go on a walk – try to keep it a calm exit.

If you are driving, do not let your puppy (or dog) jump out of the car or back in. Use a ramp or lift them out and back.

- If you are planning for your puppy to have some free running, walk them for a few minutes first to ensure they have warmed up.
- Allow them to sniff, as this also aids good body movement and tissue warming.
- Allowing to sniff will also aid calming.
- You can also walk in curves, not just straight lines.

Age: When they are immune and safe to walk and be amongst other dogs.

Frequency: Max two times a day. The amount of time to be calculated alongside other **High Impact** activities that are being conducted. Probably two high impact activities will be incorporated at the same time, so the combined time has to be adjusted.

For how long? Up to five minutes, two times a day.

Preparing your puppy: Preparing 'you' is more important. Take some treats for aiding a recall or reinforcing a good situation/behaviour.

Switch your mobile/cell phone off, you need to have **full concentration** on your puppy, on or off leash.

Bad things can happen in a split second when there is a lack of concentration.

Have lessons in lead work *so your puppy has a chance of understanding what is expected of* it and you both can reduce the possibility of injury – immediate and ongoing.

Post-exercise: Physically – to end the walk, allow some more sniffing, this is a great warm-down exercise to help regain tissue homeostasis.

Mentally – depending on how the activity progressed, the puppy may be a little excited and not wanting to relax immediately. Sniffing and a calming activity could work really well, and then, time being left for them to sleep and rest.

Equipment: A well-fitted harness that is adjusted **and upgraded** according to their growth and development.

A long leash. A leash should be at least six feet or two metres long (a retracting leash is NOT recommended as the locking device may not be triggered in time to prevent a puppy from injury; see equipment, page 207).

A good trainer will ensure that you learn how to encourage your puppy to walk safely on the lead.

Aim: To have a stimulatory outing where your puppy can learn and develop both their social as well as physical skills (Figure 5.5).

Figure 5.5 An inquisitive puppy enjoying his own exploration (not the best type of harness).

Table 5.4   High Impact: How the Exercise Will Help Build Your Puppy

How High Impact Exercise with a Well-Fitting Harness Can Be Beneficial		
Brain/neural pathways	**Skeleton**	**Muscle**
Positive socialisation experience. Develop and reinforce muscular neural pathways. Bonding experience between human and puppy. Develop olfactory sense.	Appropriate loading and stresses to help strengthen skeletal structures.	Develop and reinforce good muscular movement patterns. Develop different paces and practice movement skills.

This whole experience should be fantastic fun for both of you, so you as the human should have lessons in how you both can get the best out of this experience – both now and for the future (Table 5.4).

LIMITATIONS AND CAUTION
- Potential for overstimulation.
- Impede movement patterns.
- Change gait style from walk to trot.
- Create incorrect loading on joints and skeleton due to repetitive asymmetric actions.
- An incorrectly fitting harness with the puppy pulling is potentially damaging muscles in the neck, throat, and forelimb (see equipment, page 207).
- Asymmetric movement patterns from walking consistently on one side

It is really difficult to stop a puppy from pulling on the lead, it is, however, extremely important to receive good instruction on how it can be reduced and better still prevented. The action of a puppy pulling through a restraining device (collar or harness) is damaging to varying degrees depending on where the pressure points are on the puppy.

Whatever the equipment that is used, a puppy that is pulling will exert uneven and misplaced load through their body, creating excessive loads going through their muscles and joints, creating huge potential issues in their structure. This type of damage will not become obvious for months or even years (see page 207 on collars and harnesses for other issues pertaining to restraining equipment; Figures 5.6 and 5.7).

For a puppy that is pulling on the lead, the anatomical position of its legs is changed. This could repetitively have an impact on the loading and joints of the affected region, primarily the hip joint.

Figure 5.6 A puppy pulling on a leash, which involves splaying the hind legs to gain more traction but putting the legs in an incorrect anatomical position.

Figure 5.7 A puppy walking without pulling on the leash, maintaining good functional movement and correct anatomical posture.

FREE RUNNING

Free running is without a leash, because even if a puppy is on a loose leash, it will influence their body position when walking or running, and they will not be in an anatomical straight line. This type of minor negative repetitive movement will have a damaging effect on your puppy's gait and body symmetry (Figure 5.8).

Free running for limited periods for a puppy is important because 'free running' creates positive load upon the skeleton and joints.

WHAT IS FREE RUNNING?

See Table 5.5.

Free running off leash – only if the environment is safe for the puppy to run free without risk of dangers or potential injury.

If your environment does not offer the opportunity to free run your puppy, then a place must be found where the puppy can safely be allowed to free run.

When free running or walking, the puppy should be given the choice of when to run and importantly, **be able to stop – and rest**.

Safe free running is also important psychologically for a puppy. They can experiment with their body, they can test their speed, they can challenge themselves physically, and they can stop when they need to! (Table 5.6).

Figure 5.8 A puppy enjoying free running: self-initiated, stopping when he wishes. Expressing free will and choice gives a puppy confidence and autonomy.

Table 5.5 The Components of Free Running

Free running – off leash High impact activity (see programme)	Forced running (potentially damaging) Ultra-high impact (see programme)
Running in a garden without stimulus, such as having toys, balls, etc. thrown for them	Running in the garden being chased/played with by a larger dog
Running in a garden, playing with toys – by their own volition	Running in the garden having balls/toys thrown
Running in a garden, exploring, and sniffing	
Running on a walk without being pursued or encouraged/induced (important that they can rest when they choose) plus: • Being very close to home • Close to the car • Be **comfortably carried home** So they can rest when they choose.	Running for long periods on a walk – being chased by other adult dogs
	Running on a walk – being chased by the larger family dog
Running and playing with a toy, **self-initiated**	Running on a walk – having balls, etc. being thrown
Running around the home totally on **non-slip** flooring	Running around the home on slippery floors See page 174
	Running alongside a human as an inducement or on a leash
	Running alongside a human on a bicycle

LIMITATIONS AND CAUTION

- Free running puppies tend to self-regulate. However, when they are being allowed or encouraged to run more than the optimum time, they can become overly excited and not able to settle.
- They could continue to run without due care as they are **running on adrenaline**.
- This is a very likely issue. It is best to be prepared and have a calming activity ready for them.
- Too much running can be detrimental to skeletal development.
- If a puppy is encouraged to chase an object or person, even in a game, it can stimulate them to overexercise.

135

Table 5.6   High Impact Free Running

Activity: Free running – high impact (Figure 5.9)

Always must be conducted on a non-slip surface

Never leave your puppy unattended

Figure 5.9 Free running is so important for a puppy's physical and mental health.

Instructions: Important to **read** the information on **Free running**.

This can be done inside or outside but ensure that it is a safe place, with no unwelcome interruptions from other dogs or humans.

If your puppy is free running in the house, do not allow them to jump on and off the furniture. This would be really damaging for their bones and joints.

Age: From eight weeks around the home

then outside from 14 weeks until they can go out and socialise.

Frequency: Depending on how long they are running; puppies at this age are quite good at self-regulating **if they are not** encouraged to chase anything, like a ball or larger dog.

For how long? Two minutes × three times a day

Preparing your puppy: Prepare or adjust surroundings.

Ensure they are in safe surroundings and environment.

Post-exercise: If they are overexcited and not able to settle, offer a settling activity so they can settle and sleep.

Equipment: An open area either indoors or outdoors

It must be on a totally non-slip surface.

Aim: For your puppy to run with free choice. Puppy's need to run but to a natural proportion, so let them run a natural amount of running (using no forms of encouragement).

Progression: As your puppy grows in size and confidence, they can have incrementally more time free running.

Allow your puppy to self-regulate. If they want to have a burst of running, that is not lured, encouraged, or externally stimulated, this is positive.

If they are running and chasing another dog, and it is a happy, playful, equal in assertion, only allow

8–12 weeks: 6 mins max per day

12–16 weeks: 12 mins max per day

etc.

Table 5.6 (Continued)   How the Exercise Will Help Build Your Puppy

Brain/neural pathways	Skeleton	Muscle
Reinforces the neural pathways developed through slow movement into functional use for fast movement.	Important for them to naturally use their body, to run and assert natural loading onto the long bones, to aid positive development, aiding strength to their long bones.	Developing the power muscles range of movement.
Creates excitement (this must be restricted as they do not have the coping measures at this age to be able to self-calm).		**Develops the heart muscle.**
Allows them to practice different actions at speed.	Takes joints through their natural range, aiding joint health.	Enables practice of the running stride length.
Reflect joy, health, and well-being.	It is so important for long bone cellular development, enhancing compressive, **direct, tensile, and rotational strength.**	
Condition their muscles and joints to accept a longer range of movement.		

- Overexercise is not good for joint formation or bone formation. It is possible that over running could impact concussive load through the growth plates and newly forming cartilaginous joints (see page 78).

Too much running will unbalance the muscular system. It will promote power muscle usage and demote foundation muscles. This causes instability and imbalance in the body that can progress to power muscle domination that disconnects the functional fascial bands and hugely reduces functional movement. This causes the dogs to form new and less effective muscle patterning. With the foundation muscles being demoted, the puppy will be left without good stability within their skeleton, which can go on to form secondary issues within an unstable skeletal system. It could also leave them more open to injury due to lesser proprioception and being less agile (Table 5.7).

Table 5.7 Low Impact Exercise over Poles

Walking over 'messy' horizontal poles – low impact (Figures 5.10 and 5.11)

Must always be conducted on a non-slip surface

Never leave your puppy unattended

Figure 5.10 Outdoor over some very low obstacles (in scale with the size of the puppy).

Developing the ability
to use their legs
separately
'The Joined up
Puppy'

Figure 5.11 A puppy using natural movement patterns and gentle flexing through its whole body that helps to produce 'the joined-up puppy'.

(Continued)

Table 5.7 (Continued)   Low Impact Exercise over Poles

Instructions: Lay four to five poles on the ground in a random arrangement, so they are loosely scattered over the ground.

Throw a few treats amongst the poles to encourage your puppy to slowly walk over them.

This can also be done inside or outside but ensure that it is a safe place, with no interruptions from other dogs or humans

See Figure 5.12.

Figure 5.12 Ensure the poles they are walking over are no higher than the top of your puppy's toes.

Age: Eight weeks but ensure the equipment is the correct scale for the puppy. The pole or obstacle should be lower than the height of their foot.

Frequency: Once a day.

For how long? Not longer than five minutes.

Preparing your puppy: Ensure they have had a small amount of their normal food before they look for treats (see treat search, page 138).

Post-exercise: This is much more tiring than it looks. Allow them to relax and be uninterrupted so they can sleep. If they are a little excited, calm them down with a settling activity, see page 151.

Equipment:
- Small plastic poles
- Canes
- Plastic hoops
- Sticks
- Broom handles (when they are larger but without the broom head attached, so the pole can roll)

Progression *(as your puppy grows, in size and confidence, more challenging items can be added)*: Vertical poles could be added if the exercise is being conducted inside, and horizontal poles can be placed under the table and chairs.

Aim: This is to be conducted slowly, enabling their feet to move around the poles slowly and independently. The random pole arrangement is critical to this exercise, as this is not an exercise to establish a gait or stride length.

Also aids natural neck strengthening through extending and lifting their head and neck.

Brain	Skeleton	Muscle
This is a very powerful neurological 'pathway' exercise. This reminds your puppy of the different movement directions they performed when they were playing with their siblings.	Asserts positive, gentle, and diverse load over the joints and bones.	A very important exercise for encouraging more movement directions, which will activate the foundation muscles building your puppy's stability.
This is a great exercise for developing new and vital neurological pathways to your puppy's foundation muscles.	Promotes synovial fluid production from varied joint movements.	Encouraging isolated movement through their pelvic region and hind legs, activating their deep hip stabilisers or foundation muscles.
This develops your puppy's coordination and proprioception		Aids natural neck strengthening.
		Enabling balance through their whole body plus engagement of the tail.

Table 5.7 (Continued) How the Exercise will Help Build Your Puppy

LIMITATIONS AND CAUTION

- Add additional equipment slowly.
- Do not interfere with your puppy by encouraging or guiding them.
- Do not have anything higher than their wrist that they may jump off (wrist or carpus, see anatomy, page 31; Table 5.8).

Table 5.8 Low Impact Exercise with Poles

Low impact (Figures 5.13–5.15)

Must always be conducted on a non-slip surface

Never leave your puppy unattended

Figure 5.13 Encouraging natural movement patterns by encouraging your puppy to weave around the legs of a table and chairs on a non-slip surface.

Figure 5.14 When your puppy can cope with vertical poles, you can add some of the horizontal poles.

141

(Continued)

Table 5.8 (Continued)   Low Impact Exercise with Poles

Figure 5.15 Encouraging all forms of natural movement patterns includes going under obstacles (slowly) too.

Instructions: This can be done inside or outside. Inside furniture can be utilised really easily. **Tables and chairs** are wonderful vertical poles.

Outside, **cones and poles can** be used, or walking through **tall grass and around objects** – it can be interpreted in many ways.

This can also be done sitting on the floor, with your knees flexed and encouraging your puppy to go around and in and out of your legs.

If outside, it is better if your puppy is **off leash**, it enables better free movement and better choice of direction.

Age: Eight weeks to 18 years. The obstacles you use very much depend on the size of your puppy.

Frequency: This can be done two or three times a day. It can involve having a toy being put amongst the legs, and they must retrieve it.

For how long? This could be really quite tiring, so a couple of minutes at a time.

Preparing your puppy: If you are planning to use treats, it is advisable to give them a small part of their meal first.

Post-exercise: This is a highly stimulatory and physical exercise. After the exercise, it is important for their physical development to be allowed to rest. If necessary, apply a calming technique.

Equipment: Vital that it is performed on a non-slip surface.

Progression *(as your puppy grows in size and confidence, additional more challenging items can be added)*: This can be progressed by adding some treats amongst the chair and/or table legs, see page 138 for treat search.

Obviously, tables and chairs could become too low for larger puppies and dogs, then cones with poles would be better.

If you introduce equipment ensure it is light and the correct scale for your puppy.

Aim: This exercise is to be conducted at a walk, allowing for the puppy to gently weave and safely move and flex in both directions around the vertical obstacles.

Table 5.8 (Continued)   How the Exercise will Help Build Your Puppy

Brain/neural pathways	Skeleton	Muscle
Develops lateral and medial movement or sideways movement. This will encourage cross-lateral exercise, which is great for limb stability and brain development. Develops enhanced proprioception.	Asserts positive gentle and diverse load over the joints and bones – including their feet. Promotes synovial fluid production from varied joint movement.	Develops lateral and medial movement or sideways movement. This will encourage cross-lateral exercise, which is great for limb stability and brain development. Encouraging isolated movement through their pelvic region and hind legs, activating their deep hip stabilisers or foundation muscles. Develops flexibility through the trunk and encourages neck extension. Enhances coordination and spatial awareness. Aids balance when turning. Encourages nose to tail connection and balance.

LIMITATIONS AND CAUTION

- Add additional equipment slowly.
- Allow your puppy free choice or they could tackle something they are not prepared for.
- If integrating horizontal poles, do not have anything high or anything higher than their wrist that they may jump off (wrist or carpus, see anatomy, page 31; Table 5.9).

Table 5.9 Low Impact Exercise with a Food Search

Food search – low impact (Figures 5.16 and 5.17)
Must always be conducted on a non-slip surface
Never leave your puppy unattended

Figure 5.16 Outdoor food searches alone, on different surfaces.

Figure 5.17 The same food search but can be done inside too.

(Also known as a 'treat search', but this name can give the impression that 'treats' should be used. Some treats may not be of such good quality as their diet requires and it is imperative that good quality and puppy appropriate food be used for this activity.)

This can also be integrated within:
- Enriched environments
- Vertical poles
- Horizontal poles

(Continued)

Table 5.9 (Continued)   Low Impact Exercise with a Food Search

Instructions: This can also be done inside and outside but ensure that it is a safe place, with no interruptions from other dogs or humans.

Age: From eight weeks to 18 years.

Frequency: Can be done after each meal.

For how long? For as long as the puppy takes to search for the treat.

Preparing your puppy: First, ensure you have taken the amount for the food search out of your puppy's ration.

Find an appropriate safe place where the food search can take place. It may have small obstacles (see vertical/horizontal poles and enriched environment). Give a cue word, so your puppy starts to learn a fun and rewarding word.

Post-exercise: This should be a calming activity and will lead the young puppy to sleep and alone time. The older puppy may just need something to chew or tear to complete the scavenging process.

Equipment: Good quality food, either from their diet or from another good quality puppy age-appropriate food. It is better if the treats are small to encourage active sniffing.

A safe area where all the food scattered will be accessible, not leaving the puppy open to dangerous situations, or impossible places for the puppy to get around, under or into.

Progression (as your puppy grows in size and confidence, additional more challenging activities can be added): This can also include other dogs known by the puppy. Before doing this, ensure there is enough space for both to move around independently and enough food for both to find.

If this is conducted correctly, it can be a bonding experience for both dogs.

The pace can be slowed by adding the food in boxes or wrapped paper (see calming/other activities).

To include a safe undulating surface will add another dimension to the puppy's development. A bank or ridges will gently challenge the muscles further, naturally adding to cross-lateral, or sideways strength.

Table 5.9 (Continued)   How the Exercise will Help Build Your Puppy

Brain	Skeleton	Muscle
Fulfilling an innate act.	Asserts a positive, gentle, and diverse load over the joints and bones.	Head and neck extension facilitates the engagement of muscles and fascia through the body.
Developing their olfactory sense 'superpower' (scenting).	Promotes synovial fluid production from varied joint movement.	Encouraging isolated movement through their pelvic region and hind legs, activating their deep hip stabilisers or foundation muscles.
Facilitating good 'motor' neural activation by multidirectional movement.	Adds stability to cervical or neck vertebrae through naturally working the neck muscles.	Food searching can be slowed by putting food in boxes or paper.
Joining up their whole body from their nose to their tail, enabling enhanced proprioception.	Development of their toes and feet, strengthening soft tissue connections between their toes.	Aid natural development of neck muscles and fascial connections from their neck connecting through to their tail.

145

LIMITATIONS AND CAUTION

- Further to studies, it is recommended to give a puppy or a dog a small amount of food before they embark on a food search. This gives the digestive system something to work on to maintain good gut health.
- Add additional objects slowly.
- Do not interfere with your puppy.
- Do not add anything higher than their wrist that they may jump off (wrist or carpus, see anatomy, page 31).
- Keep watching to ensure they do not choke or eat anything inappropriate.

LIMITATIONS AND CAUTION

- Add additional equipment slowly.
- Do not interfere or encourage your puppy to move over different obstacles or challenges, leave them to tackle the obstacles they feel happy with.
- Do not have anything higher than the top of their toes, that they may jump off.
- Use equipment that is light and will move if your puppy trips.

OTHER LOW IMPACT ACTIVITIES TO AID THE DEVELOPMENT OF YOUR JOINED-UP PUPPY

WALKING ON THE SIDE OF A HILL/UNEVEN GROUND – *LOW IMPACT* – FROM EIGHT WEEKS OLD (AS LONG AS IT IS NOT TOO STEEP)

Inviting a puppy to walk on uneven ground or over different gradients is a powerful exercise that has multiple positive effects on developing their physicality.

It is important that you do not introduce any gradients that are too much for your puppy, but if you can find some undulating ground and invite them to scent, or if it is safe, include a small food search. This can really be put into functional use with the development of the horizontal and vertical poles and enriched environments.

If development work has been effective, they should, by their own volition, be able to control their body's going up, down, traversing, and turning on the gradient.

Being invited to go down the hill will help to develop their 'brakes', or how to stop. The brake muscles are the same as their accelerators, but they are applied using the muscle fibres differently, so practice with this is essential in the same environment before they need to negotiate steps or stairs.

Traversing the hill will help to develop the foundation muscles that lie within their vertebrae.

This is also good for their confidence and total coordination (Figure 5.18).

Figure 5.18 Introduce gentle gradients slowly and allow the puppy to negotiate by their own volition.

BALANCING – *LOW IMPACT*

From 12 weeks, a wide plank on the ground can be put into the enriched environment.

Leave your puppy alone to work it out for themselves.

Naturally, most dogs love to walk along logs to use their balancing skills. This is something that is not really included within a human home environment. The one time that a puppy or older dog may have to use a balance is when they are introduced to a ramp. But this might not occur until they are less mobile, and then you are asking them to learn a new physical skill.

There is another reason why this type of exercise is so good for your growing dog. It helps to develop the ventral, adductor muscles (the underneath muscles that draw the legs towards their centre line). These muscles are vital for stability and cross-lateral moving. The adductors on the hindlimbs are also vital hip stabilisers.

As long as the plank is low, if the puppy falls off/steps off the side unintentionally, it will help to learn that they can cope and recover after having a small accident.

This is a really good assessment for your puppy because if they are 'joined-up' they should be able to walk over a low balance. This exercise also teaches them how to dismount safely (Figures 5.19 and 5.20).

It is best NOT to use treats to induce them to perform these types of exercise. Just help them to understand by indicating with your open hand, guiding where you are asking them to walk.

They should be encouraged to have a low or natural head carriage to be able to engage their whole body (Figures 5.19 and 5.20).

As your puppy develops their skills, you can slightly lift one side, so they have to balance on a slight slope.

Figures 5.19 They should be encouraged to perform activities such as balancing, by invitation, but allow them time to arrange their foot placement.

Figure 5.20 Start this exercise with the plank on the ground, or if your puppy chooses to walk down a log, allow them to, without interfering.

DOWNWARD DOG – *LOW IMPACT*

This is quite an exaggerated view of this exercise, but the principle is to allow and gently encourage your puppy to *safely* look down into a shallow hole.

The aim of the activity is to practice using the foundation muscles within the pelvic region and also to extend the hamstrings. The hamstring group is a huge power muscle group, as they extend the hind leg, enabling power forwards (only

Hamstring
muscle group

Figure 5.21 When a dog is invited to lower their nose lower than their feet, as in looking into a hole, it can really help to naturally extend their major power muscles.

two of the three muscles are indicated, the third is situated just on the inner thigh). However, this group of muscles can become quite congested as within our environment there are few opportunities where dogs are invited to extend their head and noses below ground level.

This can be very effective even if the hole is very shallow (Figure 5.21).

All these games should be low impact if they are carried out without force and on a non-slip surface.

THE HUMAN ACTIVITY CENTRE: ACTIVITIES FOR YOU BOTH TO ENJOY

These games for the two of you are wonderful for interaction, the emotional bonding created from 'playing' with your puppy is enormous. Likewise, these games are great for their physical development because naturally they would develop using play with their siblings. But please remember that you should try to replicate their siblings' size and strength.

TOY PLAY

Puppy's love playing with toys but generally their favourite game is playing with toys and their guardian! Games are what develops relationships in the natural world. Playing is a vital part of our development, so playing with another being that is involved in the game is a very bonding and trust making activity.

THE HUMAN ACTIVITY CENTRE

With very young puppies, a game or interaction of any kind can often very quickly and easily descend into an unsuitable biting game. Whether they are very sharp baby teeth or more mature teeth, as a human we are not so keen on that variant!

Figure 5.22 Turn yourself into a safe, soft, and fun human activity centre for your puppy. All activities must be performed on the floor and on a non-slip surface.

To introduce a toy during or at the beginning of a game can offer a wonderful redirection from your body back onto the toy! (Figure 5.22).

Also, if you are down on the floor with them, they can use your legs as soft and low climbing equipment, which is great for their physical development. **Ensure you are on a non-slip surface**.

If you flex your knees, you can *encourage (not drag them)* them to walk gently around your static legs, and then over and around. A really challenging but totally fulfilling game for both of you. The physical and mental stimulation will satisfy your puppy after a very few minutes, and they will want to have a relaxed sleep to process and rest. (Keep the play calm and slow, and do not try to encourage your puppy to get excited by encouraging them to run and twist – that can have a detrimental effect.)

ENCOURAGE THEIR NATURAL BEHAVIOURS AND TRAITS

Playing with your puppy is also a great opportunity to find out about your puppy's breed characteristics; you can then make a game more focused towards what their breed's function was originally.

By knowing this, it is even more fun to 'play to their strengths' but also start to consider what **they** would like to do, or their **purpose**.

Dog's, just like us, need a purpose, perhaps a job or responsibility. Not too much, but enough to feel important within the household. Not just barking at the front door. This can often be misconstrued as being their purpose, but it can be fear-based rather than fulfilling for them (indeed, actually stressful).

If we give them a job, a purpose, they will be less likely to invent their own, which might be inappropriate to where you live or your environment.

TUGGING – THIS SHOULD BE CALLED 'THE GIVE AND TAKE GAME'

Puppies tend to love tugging and it is a natural and good activity for them. It can also be a bonding activity for you both to participate in.

However, there is a potential for it to have negative physical effects. It is critical that when tugging you behave as if you are one of your puppy's siblings or a gentle adult; you must suit your 'tugging' to your puppy's strength and height.

It must be remembered that puppies have very delicate teeth, and their sharpness does not replicate their strength (Figure 5.23). Puppy teeth are intended to get them going onto solid food, to get to know about bite inhibition, but they are not intended to be put under great stress. In the 'wild', puppies are still having their food regurgitated up until 15 weeks, therefore the need for tugging at food between siblings is not a driving behaviour.

It is not just their teeth that can be affected by over-zealous tugging, their jaw is also really weak, so if the tugging is too exuberant, it can affect the lie of the jaw, which could have lifelong issues (see TMJ and the jaw, page 12).

It is not intended to be a battle between you both, it is a game of 'give and take'.

Positive 'give and take'

Tugging is a great exercise for developing strength through appropriate and gentle resistance, and load. The resistance is through the amount of 'hold' you offer to resist your puppy with the object; and the load is the direction of pull to provide, in relation to their height. This is also important for a puppy's neck and jaw strength and development.

The neck is a vital anatomical structure for all mammals; almost every physical action starts with the engagement of the neck. It is a highway of information and

Figure 5.23 When puppies are biting or playing with their mouths, their teeth can easily be broken with 'tugging', especially under five months old (20 weeks).

nutrition, carried via the airway, gullet, arteries and veins, muscles, fascia, and, of course, nerves.

A puppy needs to develop a strong but flexible neck. There are lots of activities that help this, and a limited amount of appropriate 'natural' tugging is another excellent activity for your puppy to participate in to develop soft tissue strength and to help stabilise this vital structure (Figures 5.24 and 5.25).

Figure 5.24 An adult, sympathetically adjusting their resistance in relation to the size and age of the puppy and also the load or height in which they are holding on to the toy.

Figure 5.25 An adult naturally lowers themself to offer appropriate load and resistance to a young puppy.

Negative 'give and take'

If the tugging is out of alignment with the puppy's body, the load through their vertebrae, hips, and elbows can be anatomically detrimental and will not help develop the natural loading and resistance that a puppy needs.

- Potentially, if the game is too extreme, in both height and resistance, it can be highly detrimental to the joint, muscle, and fascial connections.
- Possibly dislodge the alignment of the jaw and therefore affect the tongue and ongoing balance of the puppy.
- It can break or damage the construction of the puppy's tooth and possibly affect future development.

This may give the impression of building muscles, but there is a real risk of building the wrong muscles,* and also the compaction of joints. Therefore, there is also a risk of compacting neural pathways (Figures 5.26 and 5.27).

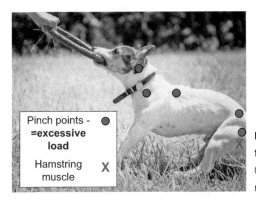

Pinch points -
=excessive
load

Hamstring   X
muscle

Figure 5.26 There is a huge difference between tugging correctly and tugging incorrectly. Correctly, can be beneficial, incorrectly is highly negative.

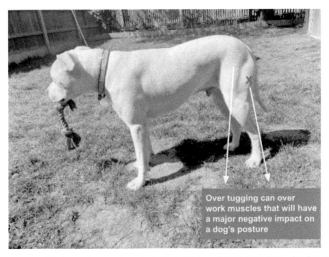

Over tugging can over work muscles that will have a major negative impact on a dog's posture

Figure 5.27 A one-year-old dog who has been involved in too much tugging – he is on high levels of pain medication.

If there is too much resistance and height or load put through the tug toy, there can be a compaction asserted through the vertebrae and massive load being put through the puppy's hindlimbs, hips, and stifles.

Like all exercise, if this happens once or twice (not a day, once or twice ever) it shouldn't be a problem, but if this is done repetitively (every day) then the ongoing negative issues can be profound. But these effects will take place insidiously and like all repetitive detrimental activities, will not be immediately obvious that harm is being done.

Why is this type of incorrectly targeted muscle development wrong?

*The muscle marked with an x is a major hamstring muscle, so important for extension of the hindlimb for all gaits but of course especially for running. This type of resistive exercise is very good for muscle bulk development, BUT it is so because it does it by breaking down microscopic muscle fibres to stimulate better and stronger development (eccentric contraction). What happens within this process is that the muscle fibres get shorter – that is why **any fitness programmes must be carefully and professionally managed**. In this situation, this dog's hamstrings could develop looking strong but actually be anatomically shorter than they are intended, which shortens the distance between the joints that the muscle is attached to. With the hamstrings, that is the pelvis and the stifle and hock. If the hamstrings shorten, it can create lumbar, hip, stifle, and hock problems. From anecdotal evidence of cases seen by Galen Myotherapists, shortened hamstrings are a major contributor to lameness and muscle pain in dogs.

HIDE AND THEN SEEK

Even breeds not necessarily known for retrieving can love 'hide and then seek' with their toys (Figure 5.28). This again is a very bonding game, and you can then marvel

Figure 5.28 Not all puppies want to play what may be thought of as breed typical games!

Figure 5.29 Play games that reflect your puppy's breed or breed type. Some breeds may be surprisingly nurturing.

at your puppy's incredible intellect and connection with you and the game. It can be the start of introducing them to a vocabulary so they can be even more connected to their human world.

The toy can be hidden on your body or within the environment. Make it easy to begin with, but when they realise how pleased you are when they play along, you can start making it a little trickier.

This game is stimulating for them and, if played to their level of comprehension, highly satisfying. Great nose work.

If they are 'seeking' in a small area, then they will be using a slow, multidirectional action that will help their stability.

DON'TS
- Only play on a non-slip surface.
- Do not include obstacles that they have to climb.
- Do not include obstacles they have to jump off.

Work with your breed type.
- What breed and therefore what inspiration for that breed (Figure 5.29).
- Also developing purpose.
- Doing things with their guardian is so rewarding and bonding for both of you, even if it is only one small activity a day.

CALMING ACTIVITIES

The turbo puppy – higher levels of physical activity can build high levels of brain activity, which your puppy may display as physical exuberance, such as:

- Racing around the home or garden.
- Displacement behaviours:
  - Jumping up.
  - Biting.
  - Destruction through excessive behaviour, chewing, or digging.
  - Frantic and almost maniacal physical activity (bouncing off furniture etc.) (Figure 5.30).

These behaviours can be driven by adrenaline, frustration caused through overexercising, and/or overstimulation, or even by underlying fatigue or overtiredness.

When a puppy becomes so animated and hyperactive, they find it difficult to settle, so an activity that enables them to naturally calm themselves is essential (Figure 5.31).

Figure 5.30 'Turbo puppy', deflecting, biting, hyperactive – but overstimulated and tired NOT needing more stimulation or exercise.

Figure 5.31 A puppy that is still active after exercise will possibly need a calming activity such a shredding (or other exercises described), this can aid calming and restorative sleep.

If your puppy returns from a walk and continues to race around, please do not think they need more exercise. They actually need to have a calming activity to help them to stop.

SNIFFING

Sniffing is so fulfilling for your puppy and also your mature dog. This previously mentioned 'superpower' is very powerful in expanding your puppy's physical and emotional development.

According to Professor Daniel Mills of the University of Lincoln, a puppy requires at least ten seconds to really absorb and process each smell; if you remove them before that time, they need to start again. Ten seconds may seem quite a long time to wait, but within that time your puppy is:

- Developing their olfactory system (scenting skills).
- Developing their brain connections to discern and differentiate different smells, both organic and inorganic.
- Catching up on what is going on in the neighbourhood – the gossip. 'Gossip is not frivolous as it sounds. Gossip has always been vital for human development and social interaction, as it gives us information about what is going on outside our own environment' (Harari, 2011).
- Moving with their head down which:

    Assists hind limb usage.

    Naturally extends their neck maintaining its flexibility.

    Allows the shoulders to have the freedom to work through their natural range.
- Moving with their head down slowly:

    Aiding their foundation muscle activation and development.

    Allowing and encouraging lateral and medial movement, or sideways movement, encouraging cross-lateral (strengthening) movement.
- It is also thought to reduce pulse rate in dogs, which would be a helpful aid to calming and relaxing (Budzinski from DogFieldStudy).
- Fulfilling for the puppy and helping to develop their very own 'superpower' (Figure 5.32).

Figure 5.32 Puppies sniffing and scenting, which helps to develop their natural instincts, exploratory as well as calming.

CHEWING

- A calming activity.
- A low impact development activity.

Chewing is such a natural activity for a puppy, and it is important that we can find suitable objects/toys/bones for them to chew on safely.

There are so many chew toys and 'chews' on the market, be careful that they are safe and will not cause damage by blocking the puppy's gut by splintering or breaking. Your puppy should not be left alone when chewing.

Chewing on a suitable raw bone

Chewing on a bone is a fulfilling activity for an **older** puppy/adolescent as well as a mature dog. This is obviously a natural way of eating for a dog and chewing and stripping of a bone also has good physical development features (Figure 5.33).

The action of holding on to the bone is so good for the dexterity of their toes. It also works on their fine motor skills, enabling their toes (anatomical human equivalent fingers) to grip and manipulate the bone into different positions. They use their dewclaws (anatomical human equivalent to thumbs), which demonstrates how dextrous and therefore how vital the dewclaw is for the dog.

It also aids a safe and natural neck strengthening exercise by stripping the bone of meat and fat. It offers the correct amount of natural resistance that is good for muscle strengthening. Dog's necks are vital for all movement and are often overloaded through incorrect exercise and activities. But this activity is natural, and aids jaw and toe (finger) strength.

Because the toes are being used in an extended and gripping position, it is a wonderful complementary exercise for a puppy or dog that does a lot of road walking; it naturally stretches the toes, giving the joints healthy 'spacing', allowing better joint health.

Generally, the shoulders are extended during chewing on a bone, which is again a good complementary exercise to counter or balance, too much impact from walking and running.

Figure 5.33 Chewing a safe bone is a natural development; it is calming and also helps develop their dexterity and exercise, as well as developing their forelimbs, feet, and toes.

Caution:

- Not advised when a puppy still has their milk teeth.
- Do not leave your puppy alone.
- Ensure the bone is safe and non-splintering.
- Ensure the puppy has a protected place for chewing; sometimes other dogs can challenge a puppy (or another dog) for a bone, whereas they possibly would not for a toy.

Not just a calming activity but also helps to develop a puppy's dexterity; using all the toes and claws, including their dewclaw, giving that an activity that will help to develop its strength (see dewclaw, page 33).

'BRAIN' GAMES

- Can be a calming activity.
- Low impact if not done repetitively.

There are lots of 'brain' games on the market and also games that can be made at home. They certainly have a place in our current environment as they can replace the natural investigations that puppies and young dogs would normally participate in (Figure 5.34).

These games can be fulfilling and good for bringing naturally outside challenges into an inside environment so puppies can hone their skills of scavenging and finding. As long as the game is not too difficult for them, they will find it very rewarding.

Figure 5.34 Brain games can be purchased or made out of pre-recycled materials.

However, if they are continuously unsuccessful, they could find it frustrating and that would potentially create difficult behaviour.

Physically, they are again moving around on the spot, therefore using slow, multidirectional movements, which are good for foundation muscle development. Ensure that these games are not overused, as they can produce repetitive movements, which can become repetitive strains.

- Possibly use the same exercise, three times a week for a maximum of 15 minutes.
- If games involve different actions, then these can be interspersed during the week.

Caution:

- Conduct on a **non-slip surface**.
- Never leave them alone.
- Monitor how long they participate, games where there can be repetitive movement using one leg if played excessively could cause repetitive strains.
- Ensure the rewards are sufficient but also not too easy.
- include this food consumption as part of their daily allowance.

CALMING ACTIVITIES: LOW IMPACT

BOXES, CARDBOARD, AND PAPER, OR PRE-RECYCLE ACTIVITIES!

These activities could potentially be the 'chewed shoe saver' (Figure 5.35)!

(Please ensure the boxes you offer are safe, with no staples, or other dangerous or toxic components.)

Boxes and/or loo roll cardboard can be a great resource for your puppy. They can be great for explorations, hiding things in, climbing over, into, and shredding. Something this simple can provide so much entertainment and good emotional and physical development.

Boxes can be used within the enriched environment or on their own. The explorations will provide multidirectional movement and slow movement but also physical challenges of climbing, and even falling over or off the box. The advantage is that they can practice their physical skills; **ensure they are in a safe environment on a good non-slip surface** (Figures 5.36–5.38).

Shredding cardboard can be so fulfilling for a puppy, as well as a release from frustration. It can potentially direct their natural shredding instincts onto something that is of no value, rather than a treasured possession.

It also helps to create natural tooth embedding to promote healthier adult teeth.

It aids the jaw and neck developer and conditioner, helping to naturally develop all the supporting muscles within these regions.

It may be messy to clear up, but this is potentially one of the best ways of allowing them an instinctive outlet, rather than this same instinct being used on something of value, like a pair of shoes.

Figure 5.35 Cardboard chewing could be messy, but it might save shoes from being chewed.

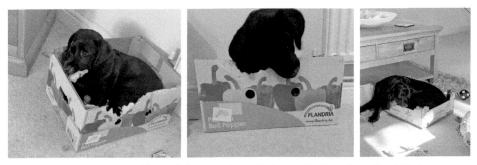

Figure 5.36 Lola – 12 weeks old has all the 'toys' but still chooses that cardboard box! Playing on a non-slip surface.

Figure 5.37 Marge loves to explore every part of the box and its contents. This is a low impact activity due to being on a non-slip surface.

Figure 5.38 The same activity but on a slippery floor will change this from a low impact activity into an ultra-high impact activity.

If a puppy is intentionally given something to shred, they can become discerning and most certainly understand what is and what isn't to be shredded, if the item is 'given' to them, and then invited to chew or shred. Encouraging words can also help. This can really help them become discerning about what they can and can't chew or shred.

Be fair to them and be clear with when they can shred, they want to be discerning, so give them the opportunity to do so.

Do not leave a puppy alone during any of these activities.

Ensure that any food given is of an equally high standard and taken from their daily amount so as not to overfeed and encourage them to gain weight (see page 95).

Natural additional activities for your puppy to enjoy their food rather than just from a bowl:

- Chewing.
- Licking.
- Searching for food.
  - Include this food as part of their daily allowance.

CALMING ACTIVITY

Give on a non-slip floor.

A 'LICKY' MAT

These come in different sizes and designs, depending on the size of your puppy. They are intended for food to be squeezed into the depressions so that the puppy (or dog) can lick and remove the food.

Figure 5.39 Licking food from a specialised toughened chew toy can be distracting and challenging but also calming.

These can be a calming mechanism for a puppy, but also useful if you just need to keep them busy for a while.

This can be very self-soothing and satisfying for a puppy, it also encourages them to extend through their body, but equally allows them to lie down if they wish; both of these are natural positions for a puppy (Figure 5.39).

TOUGHENED SPECIALISED FOOD-FILLED CHEWING TOYS

These toys can be initially quite challenging for a puppy, but they can play a wonderful role for the diversity of activity, calming and also mentally stimulating (as long as it is not too difficult for the puppy to receive the reward of food, stuffed within the toy).

This type of activity also aids the dexterity and development of their feet. It exercises their toes, especially if they have a lot of higher impact road walking, or lots of soft ground.

FOOD-FILLED CARDBOARD ROLLS

Fill an empty loo or kitchen roll cardboard with some food and then the puppy can shred and find the reward of food.

SNUFFLE MATS

These are fabric mats that are often made by threading strips of fabric through a plastic doormat to form a dense fabric nest that food can be hidden amongst. The puppy is then encouraged to 'snuffle' and find the food.

This again is a wonderful natural activity that a puppy would embark on, snuffling through the undergrowth, but instead through the fabric.

This, similar to other activities, can be physically enhancing as well as calming in a positively mentally stimulating way, satisfying and also developing their olfactory (smell) senses (Figure 5.40).

Figure 5.40 A snuffle mat on a non-slip surface is another great stimulating yet calming activity.

DEVELOPING CURIOSITY

Engaging with your puppy in everyday actions and activities is so important for their confidence and your relationship. Like us, they need to know about their environment. They want to sniff, look, listen, and maybe try out their teeth.

The more they are not included or just hear the word 'NO', or 'OFF', the more inappropriately curious they will become. The areas or objects will have a mystique that they must get to, with the possibility of them causing damage to themselves, the site, or the object.

If it is of no immediate danger, let your puppy take a supervised look in a cupboard, or in a drawer, or in a box, your laundry, anything you are interacting with, the allure will soon reduce, their natural curiosity will be healthier and more measured.

An enriched environment is a wonderful outlet for curiosity (see page 132); you do not need to add food, just interesting articles that are around the home and invite them to just sniff and explore.

Your new puppy is a sentient being that has a wonderfully absorbent 'plastic' brain that is growing, developing, and craving positive stimulation all the time. This is called neuroplasticity, as it refers to the brain's ability to change and adapt to different situations, stimulations, and experiences.

Your puppy wants to be part of your world, so invite them to be. Add the word 'trust' into your thought process; **trust them to make the correct decisions**. You may think they are being naughty, but they are just having fun. Try not to dampen that wonderful sense of joy, which will soon reduce as they mature. Embrace it, show them your world, allow them to explore, and guide them, rather than always telling them 'no'!

As a human, if we are always being told 'no', go away, we become frustrated and can either take on an introverted, disconnected persona, or one that is always pushing boundaries.

If you invite your puppy to be part of your life, allow them to look, allow them to sniff, then their desire through denial will not be so acute. They can become more self-discerning or need just a gentle reminder of what is yours and what is theirs. The saying 'you always want what you can't have' relates to puppies too!

This takes time and consistency; **puppies don't want to disrupt your life, they just want to be part of it**, so trust and invite your puppy into your world. Allow them to be discerning, this opens the most amazing potential for a deeper, more fulfilling, and inclusive companionship for all concerned.

Anecdote from the author: Tilly was nine months old at Christmas, she had never seen a Christmas tree or the adornments. I got out the Christmas tree and decorations and thought that I would have to put up a guard around it, as it was situated very near to her bed. Instead, she indicated by sniffing the decorations that she wanted to know what was going on. So we decorated the tree together (not literally, but she was with me), and she loved it. She sniffed every decoration, as if she had taken pride in how and where 'she' placed them. For the entire three weeks the Christmas tree was up, she did not touch one decoration (she did sniff them every day though!). (See Figure 5.41 on next page.)

Figure 5.41 Tilly aged ten months (left) next to 'her' Christmas tree (Maggie aged ten years on Tilly's left) next to the Christmas tree that Tilly was part of decorating.

KEY POINTS

- There is a limited period of time for muscle patterning and muscle activation to be established in a puppy.
- Natural exercises are to be encouraged and can be done inside or outside.
- Balance and variety of activities to develop different planes of movement.
- An active puppy could be an overstimulated puppy, so use calming activities.

6   *Your new puppy*

Understanding the importance of a good start and preparing for their new home

PART A: EARLY PUPPY DEVELOPMENT

It is so exciting getting a puppy – a new member of the family. Finding a good breeder so that you can share your life with a dog that has been given the best possible start in life is key, but it is difficult to know how to achieve that. There are many different types of breeders, but the best ones are those that want to produce great puppies. Breeders who are interested in the puppies and not the remuneration! Puppy owners do not need to know *everything*, but it can be helpful to know what questions to ask, and that is what this section is intended to help with.

There are so many facets to breeding and this book is not going into those details. Instead, this book aims to help you recognise what 'physical' start your puppy has had. Then, wherever you do get your puppy from, whatever their background and breeding, we provide a development programme that will enhance their physical health.

We expect that when we get a puppy, they are 100 per cent healthy and physically aligned, but so often this is not the case. Puppies, like babies, can suffer trauma during birth. Then of course, for a puppy, life in a whelping box, even though intended to protect them, can inflict injury. There are so many potential areas for injury, such as being trodden on, squashed, being adventurous, climbing, and falling from their enclosure.

One of the most common is where the puppy is picked up and then dropped because they wriggled (Figure 6.1).

Young puppies should *not* be picked up by unqualified people. It should only be done by the breeder, or someone being given permission by the breeder, because they are performing an organised procedure that is planned and managed. **Puppies do not bounce!**

If they are picked up incorrectly, apart from the risk of them being dropped, incorrect forces can be asserted on their skeleton, such as someone picking up a puppy and placing their hand between the puppy's forelimbs (Figure 6.2).

Try to pick up your puppy or mature dog as little as possible, unless they are in a dangerous position. To be picked up constantly could remove their sense of purpose and autonomy (Figure 6.3).

They will have less choice in the direction they take, which is a disadvantage they have over their larger cousins.

DOI: 10.1201/9781003268789-6

Figure 6.1 A puppy being held in a way that is putting a strain on its anatomical structure.

Figure 6.2 A puppy being held, having his forelimbs splayed, asserting incorrect forces through its developing stabilising muscles and attachments of the forelimbs.

BIRTH

The development of your puppy starts straight away, even before they are born. Is the mother happy and as stress-free as possible? Are the breeding human family prepared and prioritising care for the expectant mother and future well-being of her offspring? (Figures 6.4 and 6.5).

As the puppies get active, it is critical that they can grip and walk easily. This will develop their structure correctly from the very beginning of their lives. Vinyl floors

Figure 6.3 Puppies and small dogs do not always want to be picked up.

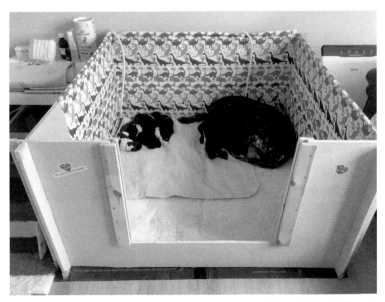

Figure 6.4 This whelping box has no pig rails. This is deliberate, as this particular bitch got very stressed with them and couldn't stretch out; however, when the puppies were born, she was incredibly careful where she positioned herself; all the puppies were safe, and none were injured.

Figure 6.5 Safe toys introduced very early for puppies to snuggle into, as a variety of toys, including those that are hanging up, are important for their development and future socialisation.

may be easier to clean, but they will not provide traction for the young pups to walk and move around (see slippery floors, page 174).

It is important that the newly born puppies receive stimulation both physically and psychologically. At around three weeks of age, these puppies are moved out of the whelping box into a puppy pen. These have hospital pads for their toilet area that give great traction for walking on. They are washable so do not create a lot of rubbish and the pups tend not to shred and eat them like disposable ones (Figures 6.6 and 6.7).

Figure 6.6 The inside of the pen barrier has Perspex to protect the puppies' feet and claws from getting caught; the floor of the pen is non-slip.

Figure 6.7 Perspex inside the pen barrier protects the puppies' feet and claws from getting caught; the base of the pen should be non-slip.

When the puppies are old enough to go outside, it is ideal for them to go onto a natural surface that offers them the greatest traction. Their physical and psychological development is supported by adding more toys, activities, surfaces, and safe challenges.

Different surfaces and levels challenge their balance and proprioception, this is a great way of activating natural foundation muscle development (Figures 6.8–6.12).

Figure 6.8 Add different types of physical safe challenges, with bright colours and large graphics.

Figure 6.9 The pink seat is a musical chair that plays a tune when they sit on it. Also, there are buttons they can press for other noises. Climbing on and off has to be done carefully and slowly, stimulating foundation muscle development.

Figure 6.10 Early introduction to different animals and situations will help with ongoing socialisation.

Figure 6.11 Handling of puppies must be conducted at ground level.

Figure 6.12 Playing and interacting with puppies must be conducted at ground level.

PREPARING FOR YOUR NEW PUPPY'S ARRIVAL

PREPARING YOUR OTHER DOG/S

Dogs like to know what is going on, just like us. They like to be kept informed about any comings or goings within their family unit. You would not invite another new family member into your life and your home without having a bit of a discussion. Especially when that new 'person' will be living with you, eating with you, and of course sharing you! And yet we expect our dogs to just accept our choices. We may want a new puppy, but the least we can do is inform our other dogs and allow them a brief introduction before they arrive.

In the natural world, animals often rely on a pre-physical introduction by way of scent.

Because we do not have the 'scenting' superpower that dogs have, we may overlook how important it is for them to have a scent introduction. This could not be easier to do before you bring your puppy home. Try and arrange a swapping of dog (and environment) scents with your breeder (or rehoming/shelter) for a pre-introduction.

At least then they have 'met' them and can gain information about the individual you are bringing into their home and environment.

ANECDOTE

One week before Tilly arrived, I wiped one of my t-shirts all over Maggie (my ten-year-old), placed it in a bag, and posted it to my breeder, along with a 'clean' t-shirt in another bag. I sent it with a stamped addressed envelope. My breeder put my scent drenched t-shirt in Tilly's bed near her and wiped the clean one over her, replaced it in the bag, and sent it back to me.

When it arrived back with me, I carefully took the t-shirt out and placed it on the ground (Figure 6.13).

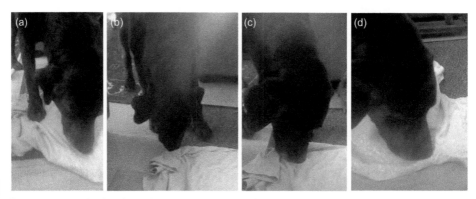

Figure 6.13 Maggie absorbing the scent from the t-shirt that was saturated in Tilly's scent.

Immediately she literally sank her nose into the t-shirt and sniffed, often with her nose perpendicular to the ground, digging her nose into the fabric, to get every last scent molecule from the fabric, 'read every line', see every 'picture', and she did this without stopping for one full minute.

When Tilly arrived, it appeared to me that they were carrying on a conversation, not starting a new one.

From the very first moment they met (I had Maggie's harness on as a safety precaution), Maggie appears to be happy to be sniffed but needed 'thinking' time to perhaps correlate this puppy with the scent on the t-shirt – we all need thinking time! (Figures 6.14–6.20).

Figure 6.14 The very first time Maggie and Tilly met, the next pictures follow in a close sequence of events.

Figure 6.15 Face-to-face sniffing.

Figure 6.16 Continued sniffing from both, Tilly sniffing Maggie's face, Maggie turning away slightly.

Figure 6.17 Tilly continuing to sniff Maggie's face, Maggie turning in more towards Tilly.

Figure 6.18 Tilly insisting on sniffing Maggie's mouth with Maggie keeping her head lowered.

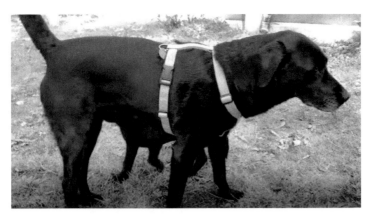

Figure 6.19 Tilly moves around Maggie.

Figure 6.20 Within 30 minutes of meeting, they were having a sniff around together, Maggie showing Tilly her new home.

KEY POINTS

- When looking to get a puppy, a breeder should be providing the best start and environment as possible.
- Prepare your new puppy using their superpower by introducing 'family' smells as a pre-introduction.
- Pre-introduce your dog and your new puppy.
- Remember your puppy will be taken from everything they have ever known – help them in whatever way you can!

PART B: PREPARING THE HOME ENVIRONMENT

We expect our puppies to be adaptable to our environment but, the sad fact is, anatomically they are not built to live in the type of homes that we do. Everything is designed for two-legged beings, at their average height and for their convenience and ease of lifestyle. Many of these factors are directly opposed to what would be suitable for a puppy and a mature dog.

Flooring

Much has been spoken about flooring and how detrimental slippery floors are generally, but especially for our puppies. Many floor coverings now are based on engineered wood, laminate, stone, or something akin such as marble. It was not that long ago that many Western homes had wall-to-wall carpets. This was much more 'dog friendly' as their feet and claws maintained good traction even when they were chasing around and running to the front door to bark.

These non-carpet floorings are popular for many reasons, but of course one is that they are easy to clean, and they fit with the new style of home. However, dogs' pads do not grip the surface, and then when they protract (extend) their claws, as they naturally would to gain more traction, in fact they gain less and slide even more.

Often, it is viewed as funny to watch a puppy sliding around like it is on an ice rink. Would it be so funny if people watched babies and toddlers falling over all the time in their own homes? The child would eventually lose confidence, be in pain, and more than likely suffer ongoing injury into their later life. This is what can happen to puppies when they slide on the floor (Figure 6.21).

Figure 6.21 Would we expect a child to live on a perpetual ice rink? Of course not, but this is what it must feel like for a puppy to live on floors that offer no traction for their paw anatomy and construction.

A dog's foot and pad are anatomically designed to maintain grip and traction on natural surfaces. Through utilising the roughness of the paw's surface, together with the unified protraction and retraction of all their claws (including the front dewclaws), they use these anatomical features to maintain incredible grip and purchase on natural terrain (Figure 6.22).

Like so many activities, it is not necessarily the one-off accident that will cause ongoing problems. It is the repetitive nature that creates these insidious damaging changes in the body, accumulating over time. Starting with puppyhood and then exponentially and continually damaging a dog's integral stabilising mechanisms will eventually expose their joints to overloading and wear.

These two pictures of puppies moving over a slippery floor show both fore and hind limbs are being splayed or abducted as they slide on the slippery surface (abduction = taken away from the body) (Figures 6.23 and 6.24).

Figure 6.22 A dog foot is not designed to enable traction on smooth flooring

Figure 6.23 A demonstration of what can happen to a dog's body when its legs lose traction on slippery flooring.

Figure 6.24 Puppies 'abducting' their legs involuntarily, the type of action that can easily impact injury through all the soft tissue and muscles as well as developing joints.

A great part of the Galen puppy plan is to develop these vital stabilising muscles. Abduction injuries, especially on puppies, can be devastating **and this activity directly damages them.** Ultimately, they are unable to function to maintain stability through their legs (see developing stability, ventral stability, page 84).

The muscles that are situated underneath the body are representative of 50 per cent of the stabilising mechanisms that maintain dogs' balance and musculoskeletal integrity for all four of the limbs; if these muscles are damaged, the dogs' limbs and stability will be compromised (Figure 6.25).

Dogs with damaged muscles will then typically start to compensate and reload their body, leading to catastrophic changes within their muscular, fascial, and skeletal structures. Ultimately, this will also negatively affect neural responses, and that will have an overall effect on the dog's health and well-being.

Smooth floors may look great, but they need to be given a non-slip rug for your puppy to follow. You do not necessarily need to cover the whole floor, but the dog will need to have a wide enough pathway to areas they need to travel to, at the entrance and exit of their beds, and also where they eat.

Maintaining a dog's toenail length is also vital, as long toenails can add to the lack of traction, especially if they are concerned and protract their nails, which basically forms ice skates under their feet (Figure 6.26).

There are various products on the market to help puppies and dogs maintain stability on slippery floors, the author's view is that non-slip rugs or carpeting is best.

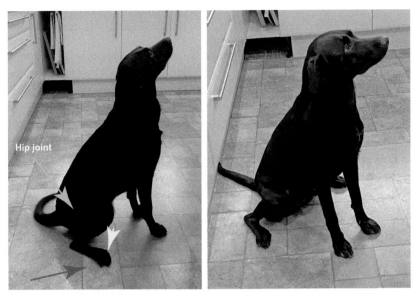

Figure 6.25 Even when a dog is sitting, slippery floors are distorting its anatomy; by trying to gain purchase on the floor, the dog is having to re-angle the foot, which is pushing the hip joint out of alignment. The dog will then have to push an incorrectly directed force through the joints to enable it to stand.

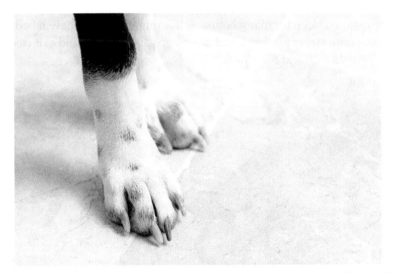

Figure 6.26 Maintain nails to their correct length. These nails are too long, especially on a slippery floor.

FEED AND WATER BOWLS

Put your puppy's water and food bowl on a non-slip rug that is large enough for the puppy, and then mature dog, to be able to stand all their legs on the rug. That way they can then eat comfortably and without stress (Figure 6.27). When a dog is eating their posture should look natural and comfortable; the legs evenly balanced and supporting the back. The puppy should not have to fix itself into a position so that they are stable enough to eat. Apart from it being detrimental from a postural perspective, it can create stress in the puppy before and during eating that could alter the gut physiology (see sympathetic nervous system, stress).

BED OR SLEEPING AREA

Where your puppy is going to sleep and feel safe is critical. There are so many different beds on the market and like us, each dog will have their own preference.

Whichever one you choose, try to find one that is supportive to your puppy. As they grow, they will want a bed that is big enough, ideally without solid sides, so they can lie on the edge comfortably. Even if a bed has a soft moulded edge, most dogs prefer to also have a movable cushion so that they can adjust the padding to make themselves even more comfortable, especially if they have an uncomfortable neck.

A puppy should sleep for many hours, so it is important they have a bed that is big enough, comfortable for them, and has a non-slip entrance and exit (see sleep, page 116; Figures 6.28 and 6.29; Table 5.1).

Figure 6.27 The posture of the puppy when he is fed off a surface where he can maintain traction is clearly different compared with his posture when on a surface where he cannot maintain traction.

Figure 6.28 There are many different styles of bed; the correct one is the one that your puppy is comfortable in.

Figure 6.29 Dogs tend to like a pillow to rest their head and neck on – just like us! Ensure their entry and exit is onto a non-slip surface.

CRATES: PRISON OR SANCTUARY?

Recently, crates have become almost 'normalised' for puppy care. Great consideration should be given before using such equipment. Remember, your puppy has just arrived at its new home and is lacking the comfort of its litter mates. Then you put them in a crate, which they may view as a 'prison' (Figures 6.30 and 6.31).

Figure 6.30 Puppies in a crate. How do they perceive the experience?

Figure 6.31 If you decide to crate your puppy, perhaps empathise with how you would feel with highly restricted movement.

You may feel that when the puppy first arrives, it is safer for them to be near you in a crate until everybody has gotten used to their new environment. This can be successful, but only with careful management. If you use a crate or playpen, then remember it should be used as a sanctuary **not a prison or punishment**. If you *must* use one to keep your puppy safe for a temporary period of time (minutes), then try and position the crate so that they can see you and offer them a calming activity (see calming activities, page 151).

If you put your puppy in a crate after an activity, you should never leave them shut in beyond their time of being happy. They can very quickly learn frustration.

Physically, crating can also impact negatively; if they are not able to stretch out, repetitively jump up trying to get attention, or if their physical activity is reduced. All of these actions can have highly detrimental effects on their development (see skeletal development, page 78).

Crates should *not* be used for your convenience. They remove a dog's choice and, just like you when you were a child and were kept contained, even if you had a room full of toys, being made to remain in one room would create stress and frustration. These emotions can be replicated by your puppy and could potentially erupt into unsociable behaviours. These can then escalate, invoking longer periods of containment for your puppy, exacerbating and creating a potential long-term problem that shouldn't have existed in the first place!

This type of lack of choice and agency can create emotional distance between the puppy and their human family. They can become physically and mentally repressed.

From a physical perspective, a lack of physical activity and freedom will have a detrimental impact on muscle and joint development.

Here are just some of the physical issues that can affect a crated puppy:

- Repetitive strains from activities produced from within the crate i.e., jumping up against the sides to attract attention, circling one way.
- Not having the opportunity through physical movement to assert correct physical loading through their bones and joints promoting strengthening.
- Not having the opportunity through free physical movement to assert correct activation to foundation muscles that support their physical frame.
- Reduced opportunity to sleep in different positions.
- Reduced movement.

The list goes on…

PLAYPENS

These offer a better space for a puppy to walk around and they provide more opportunities for them to play with toys or activities or have time away from other members of the family (other dogs or children). However, you should not leave your puppy unattended as they are not always 100 per cent secure and can collapse.

Playpens can be a better piece of integration equipment, giving the puppy needing 'time out' more space, or being removed from a situation that could cause them injury, such as being under their family's feet during a particularly large volume of feet in a reduced area (Figures 6.32 and 6.33).

However, this is still removing their free movement and free choice. Both of these elements do have an impact on your puppy's development.

Remember this type of equipment also has the capability of trapping paws and claws within the structures, as well as being toppled over by the puppy. This can cause injury to your puppy, especially if you are not there immediately to recover the situation.

Figure 6.32 Tilly in her playpen, non-slip floor with the door open. The door was never shut.

Figure 6.33 Tilly in her playpen (still with the door open) just allowing the mature dogs some 'time out'.

RAMPS AND THEIR USE

Ramps are much more accepted as part of a canine's equipment, which is a great addition to help dogs cope with our environment and the unnaturally repetitive requirements that are expected of them, such as getting out of a car (especially now that our cars, or 4 × 4-wheel drive SUVs, are generally much higher off the ground that the older styles) (Figure 6.34; see repetitive strains, page 190).

Figure 6.34 Ensure the ramp is strong enough to carry your dog. Yellow stripes on the ramp can help a dog to properly see the ramp.

Jumping out of cars regularly is seriously detrimental to a puppy's and dog's health. The use of a ramp or something similar will dramatically reduce overloading of your puppy's and then your grown dog's muscles and joints.

The type of ramp to get really depends on what you want it for, the size your puppy will grow to, and how much storage room you have in your car to travel with it.

There are 'steps' that are also available, these do not eliminate the gravitational load that a ramp would, but they tend to be smaller, so easier to store when not being used; they are better than letting your dog jump from the car! However, steps are not suitable for a small puppy, only when they are mature. Ramps are the best for puppies.

Ramps can also be useful for dogs jumping on and off furniture. Or if you have a small number of steps or stairs that your puppy will have to use regularly (Figure 6.35).

Whatever a puppy is climbing onto and off, they should never be landing on a slippery floor, and better still they should be encouraged to use a ramp. If a ramp

Figure 6.35 Ramps are great for exiting furniture, the puppy must land on a non-slip surface.

Figure 6.36 Ramps can also be used for avoiding some stairs. It is a good idea to introduce the dog to ramps when they are very young.

is introduced early, a puppy learns how to use it and becomes familiar with the balancing process: it becomes a natural movement and action (Figure 6.36).

Often, we do not introduce these pieces of equipment until our dogs are physically unstable. To then ask them to walk down a ramp that they are not 'conditioned' to do can be really challenging and distressing for these less mobile dogs, often causing them to make extreme diversions, such as leaping off the object, to avoid the ramp!

Allow your puppy to get used to ramps when they are young. Help them down the ramp and do not leave them unaided. Sometimes the application of yellow stripes can help your dog to differentiate between the ramp and the ground.

KEY POINTS

- **Prevention is better than cure** – environmental repetitive strains, especially slippery floors are highly destructive to a puppy.
- Think how you would feel if you were put in the same situation as you are placing your puppy.
- Our homes are built for humans with long legs and good traction for their feet.
- Humans and dogs are sentient beings, we feel and have emotions.
- Establishing habits to protect your puppy will be the best investment for their future health.

7   *What* not *to do! Unsuitable activities and why*

W hen building a healthy puppy, it is much more about knowing what *not* to do.

OVERWALKING YOUR PUPPY (ONE OF THE MOST COMMON REPETITIVE STRAINS)

This is probably one of the most common misinterpretations people make when they have a puppy and one that negatively affects our puppies' health. So many puppies are just overwalked (Figure 7.1). There are many reasons this happens, but one of the most common reasons I hear is, 'I need to wear my puppy out so he/she sleeps'.

There is so much wrong with that statement. First, is having a puppy a chore or a pleasure? Why do we want them to sleep all the time? Yes, puppies do need lots of good quality sleep, but as a result of great activity and stimulation, not exhaustion and overstimulation (see sleep page 116 and Table 5.1).

Everything about a puppy's body is new and developing. The cells are crazily trying to reproduce to form a larger, stronger, and ultimately healthy body. If the body is overstimulated with demands on its physiology, greater than it is designed to cope with, the pattern of growth and development is going to be disturbed.

When a puppy is overexercised, the musculoskeletal aspect of the dog's gross anatomy will respond physiologically (cellular response) to where there are greatest demands. One such area will be the puppy's joints. The joints will not be supported by the correct or sufficiently developed muscle, therefore they will be vulnerable and lacking intrinsic and vital stability (see growth plates, joints in anatomy, page 78).

The bone- and joint-supporting foundation muscles need 'slow demand' activation, with the body in the correct postural position. What this means is the head being held in the puppy's natural head and neck position: generally lower when a puppy is ambling or walking, as opposed to trotting, which is common when they are being walked.

Also, the puppy should be walking or moving slowly within different planes of movement. With extended walking in one plane of movement (forwards not sideways), it will direct the load on the same surfaces (joints and growth plates) and through unidirectional myofascial planes.

This type of activity (especially in a puppy and developing adolescent) creates a body that is overdeveloped through one plane of movement and totally undeveloped through others that are critical for their total stability, strength, and flexibility.

These overdeveloped areas of the puppy's growing body then become overloaded as progressive extended exercise continues through the puppy's development. Parts

DOI: 10.1201/9781003268789-7

Figure 7.1 A puppy will not necessarily let you know when they are tired; they just keep going (maybe they think they will be left behind if they stop?).

of their body start to wear, they 'load shift' (compensate) more into a change of posture, exacerbating loading. This is a highly destructive cycle (see stabiliser bike, page 89; wooden block anatomy, page 190; see compensation cycle, page xix; see Galen plan, Chapter 4).

As with most things, what is bad for the body is bad for the mind. The awful irony of this is that overexercising creates an overstimulated puppy, who returns from a walk that is too long and behaves as if they need more exercise. Whereas, in fact, mentally they are overstimulated and cannot settle (see turbo puppy, page 111).

THE 'WEEKEND WARRIOR'

The commonly coined phrase 'weekend warrior' refers to someone that takes their dog for short walks during the week and then very long walks at the weekend, when they have more time (Figure 7.2). For an adult this can be highly detrimental, as the dog is not conditioned to walk that distance. A bit like an individual that may run 2k in the week and then *must* run a marathon at the weekend. It is not just fitness (cardiovascular) levels, it is also strength for the muscles so that they can continue working for a prolonged period.

What can happen in this scenario is the dog returns and is very quiet for a few days, as they are trying to recover. However, they are trying to recover from damaged muscles without rehabilitation. Therefore, their muscles 'heal' but they deposit scar tissue, and the muscles get shortened, rather than stronger. This type

Figure 7.2 'Weekend warriors'. It is not good for any size or age of dog to have short walks in the week and then long walks at the weekend.

of muscle change can have large implications for their posture, which can then have secondary issues for their functional anatomy and mobility.

REPETITIVE ACTIVITIES

Most one-off activities, such as one ball throw, will probably do little harm, but it is the repetitiveness of an action that creates problems, and the type of problems that are often not seen for months or even years.

This is because the body is capable of absorbing stress, then compensating through postural adjustment, and repairing this until the body breaks down. This can often appear to happen 'overnight' or after a walk, even if they have been on the walk many times before. The reason is that the body is a little like a stack of blocks; it behaves very similarly to the game where one must take away as many of the standing blocks before the tower collapses. One moment it is standing, the next it has collapsed (Figure 7.3).

This is what happens to the dog's body. Their body can absorb so much and then it just breaks down! It starts when they are a puppy, lacking the stability, then adding repeated actions that form a multitude of repetitive strains and injuries, until they cannot adapt anymore. They are completely 'uneven' in their gait, as they are often technically lame on *all* their limbs!

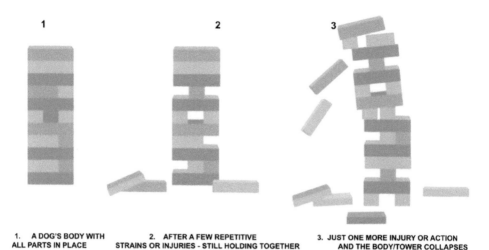

1   **2**   **3**

1.   A DOG'S BODY WITH   2.   AFTER A FEW REPETITIVE   3.   JUST ONE MORE INJURY OR ACTION
ALL PARTS IN PLACE   STRAINS OR INJURIES - STILL HOLDING TOGETHER   AND THE BODY/TOWER COLLAPSES

Figure 7.3 A wooden block game can help to represent how repetitive strain injuries slowly remove working parts of the puppy's or dog's body until eventually the body cannot function.

Repetitive strains in dogs are common and they are caused by so many different activities. The big problem is that they are often not identified until the muscular strain has made a big impact on their mobility. Even worse, they perhaps become complicit with joint disease, such as osteoarthritis, which is almost like a global plague within our canines. Osteoarthritis can also be called 'secondary joint disease': muscular dysfunction through often repeating, repetitive actions, causing repetitive strains.

Damaged muscles that are not correctly identified, treated, and then rehabilitated form what could be akin to scar tissue binding the fibres. This 'scar' tissue is inflexible, and it is complicit with the subsequent shortening of the total muscle's length. This can cause the adjoining joints (where the tendons from the muscles attach to the bones) to be drawn closer together (see muscles, page 53).

When two moving surfaces of an articulating joint are drawn closer together, joint lubricant (synovial fluid) is present within the limb joints. Continual load and stress can damage the joint capsule and therefore the integrity of this joint 'oil'. In the same way as the molecular deconstruction of old oil in a car (Figure 7.4), friction within the moving surfaces of the joint can begin, which could be a precursor to osteoarthritis, but not before painful inflammatory changes (see joints, page 49).

The joints between the vertebrae, protecting the central nervous system, do not contain synovial fluid. They have intervertebral discs that act like cushions between the joints, but they can only take so much load and strain. Similar to jumping on a cushion stuffed with feathers repetitively, the cushion will eventually burst (Figure 7.5), the feathers will not be contained, and the cushioning will be reduced. This is what can happen to the discs of the vertebrae (see intervertebral joints, Chapter 2 page 14; Figure 7.6).

Figure 7.4 All moving parts need good quality lubrication – a car needs oil and articulating joints need synovial fluid.

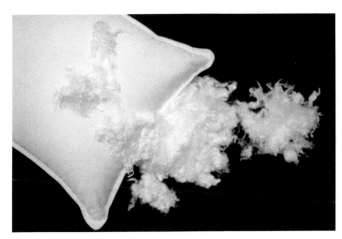

Figure 7.5 If a cushion is jumped on enough times it will burst! In a similar way to how a disc can herniate or bulge within the vertebral column.

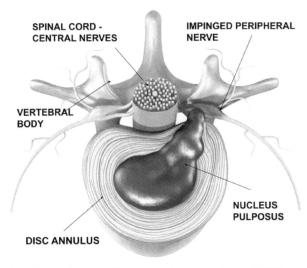

SPINAL CORD - CENTRAL NERVES

IMPINGED PERIPHERAL NERVE

VERTEBRAL BODY

NUCLEUS PULPOSUS

DISC ANNULUS

Figure 7.6 A herniated disc can be caused by repeated compression and force.

It could be argued that all of these activities are natural for a dog. Or, even more commonly regaled, 'but they love chasing/catching a ball', then 'I can't go on a walk without taking a ball' or 'they love it so much they are obsessed!' The problem is that the 'obsession' is driven by adrenaline and adrenaline masks pain or discomfort to a very high level (Figure 7.7). The fact is, *dogs do not perceive the real level of destruction being done to their bodies until they cannot easily compensate for their discomfort, and they are forced to limp or heavily adjust their gait.*

Consider the effects on both the intervertebral discs and joints of the limbs if this action, or similar, is repeated and repeated.

We often think our dogs are slowing down when they are eight years old, but that is *not* old. Maybe it is the result of a short lifetime of constant repetitive strains that have developed into secondary painful postural (muscular) and/or clinical conditions (Figure 7.8).

These long-term issues often start when they are puppies!

Figure 7.7 What starts as a highly impressive show of athleticism often ends in painful mobility disorders.

Figure 7.8 From a healthy dog full of bounce and happiness to a dog that is living with constant discomfort and possibly even pain.

BALL AND FRISBEE THROWING

In over 20 years of experience, in my opinion, some of the worst repetitive strains and body compensatory issues are caused by the repetitive activity of running after and fetching a ball. The pain from years of this activity has got to negate any 'fun' they received from the activity. Let it be clear, it is the repetitive nature of the activity, not necessarily the activity itself.

There are two aspects to this activity that can be detrimental (Figure 7.9):

1. The twisting in mid-air.
2. The landing.

Newton's third law of motion is 'each action has an equal and opposite reaction'. Interpreted in this situation is that each landing must have an effect on the body. A landing being the action, the reaction is the concussive, compressional effect on the body.

The dog's anatomy is designed to jump and land, but not repetitively, or the intrinsic structures, designed to absorb this load, become damaged and break down (see anatomy of the shoulder, page 36, discs/vertebrae etc.; Figures 7.10 and 7.11 and Table 7.1).

Figure 7.9 The extreme loads put through a dog's whole body with these types of repeated activities ultimately reflect negatively, until their 'wooden blocks' collapse.

Figure 7.10 The extreme load points as a dog lands.

Figure 7.11 Potential issue for a dog who continually lands from catching a ball or frisbee, jumps off furniture, or off high objects such as a car (see ramps, page 183).

RUNNING OR CYCLING WITH YOUR DOG

If you wish to go for a ten-mile hike or a five-mile road run, then these types of exercises are not suitable for your mature dog and most certainly not for your puppy or adolescent dog (see road running, page 197). Most breeds or breed types are not designed to run long distances. Most are anatomically and physiologically (conformation and muscle fibre type) designed for short energy bursts, i.e., sprinters not long-distance runners.

Table 7.1 shows just some of the anatomical features that will become very quickly overloaded and damaged by continual jumping and landing.

Table 7.1 Damage by Jumping and Landing	
Main anatomical areas of issue	**Effect**
1. Dynamic load being driven through the vertebrae (see Figure 7.10).	Gravitational forces driving through the intervertebral discs. The discs between the vertebrae are used to continually cushion the forces from landing (together with the potential twisting in mid-air).
2. The load being received by the muscles of the shoulders, neck, and chest. 	The muscles of the shoulders are designed for flexibility and concussion absorption, but not continual forces and loads. These become damaged and then compensate through potentially the less uncomfortable leg, then through the neck. A destructive cycle of compensation.
3. The lower neck receiving continuous load. 	The lower neck of the canine is vulnerable and has great potential for disease and intervertebral issues, which could lead to progressive neurological issues.

(*Continued*)

Table 7.1 (Continued)   Damage by Jumping and Landing

Main anatomical areas of issue	Effect
4. Muscle damage and/or compensation, postural change issues, leading to over recruitment of muscles of the head and jaw. 	This will also affect the positioning of the tongue, which can have a systemic effect on the body's balance and health.
5. The carpus (wrist) of the dog having to hyperextend (overextension) to ease the load going through the shoulders and neck. 	The carpal should be a flexible joint, allowing concussion absorbing qualities away from more vulnerable sites (such as the vertebrae). If these are uncomfortable due to overuse, the load will shift into the shoulders and neck, creating exponential issues.
6. The pelvic and lumbar region. 	Even though the load is being driven towards the neck and shoulders, the pelvic region still must land, and the landing can be heavy. If the pelvic and lumbar region of the dog is uncomfortable, then they compensate by using their forelimbs to pull themselves forwards. If, however, there are excessive loads from continual landing, this process of compensation is also going to be exponential!

See page 30 for the development and construction of shoulders, forelimbs, neck and head, and jaw.

Figure 7.12 Dogs and puppies do not share the same desire, both physical and mental, to go for extended runs without stopping.

Another reason is that dogs do not have well-adapted cooling mechanisms. They do not sweat like us: they pant but they also lose heat through their skin via their peripheral blood flow.

Some breeds, like sled dogs or carriage dogs, have been bred for long distances. However, the people who professionally run these dogs train them to a balanced conditioning programme. I would 'hope' that they would condition their dogs for the type of conditions and the type of running they would perform. Just as a long-distance runner doesn't just do long runs to train, they follow a balanced programme to ensure their body is ready for such conditions and activity.

A professional would not run a dog, with no preparation, at one speed, at one range and plane of movement, i.e., jogging, going forwards. They will most certainly ensure that they are warmed up and warmed down after an event.

Many humans like to run and think that while they are running their dog would like to come too! But this is not necessarily compatible with your dog (Figure 7.12).

When you run with your dog (even if they are off-leash), it is generally on 'human' terms:

- Where you go.
- How fast you go.
- How far you go.
- When you stop (if you stop).
- How far and where you run is very much based on how the human feels, i.e., tired, thirsty, feet hurting, muscles hurting, joints aching. The dog rarely has a choice.

Consider how you would feel if all those choices with extreme exercise were taken from you.

Figure 7.13 When running on a leash alongside a bicycle, dogs cannot choose when to stop! There is also an inherent danger with running next to a moving vehicle.

A dog loves to explore, sniff, look, listen, and then sniff again! That is natural for a dog; they love to run, but 'free running' is liberating and fulfilling; forced running can be uncomfortable and (like all activities) if done repetitively, will create compensatory issues (Figure 7.13).

They will adapt their gait pattern to the running speed and stride length that is expected of them, or the humans. This could be slower than their own natural pace, or even faster!

Dogs, like us, have their own natural stride length within each gait. A good activity regime that encourages different gaits at different speeds will condition the muscle fibres and facial connections. They will then have the ability to vary their fibre length to accommodate the different speeds, maintaining flexible muscles and soft tissue.

This type of **muscle conditioning** maintains good muscle fibre integrity that can alter according to the activity the dog is participating in.

However, if most of the exercise is done at one pace, one stride length, and one speed, this muscle integrity will be compromised. The muscle fibres will change physiologically to be stronger at the dominant pace due to the excessive loads in one position. A dog will then find it difficult to alter stride length and possibly even gait.

This can be detrimental to a dog that suddenly tries to expand its stride length, as the muscles may look well developed but they will be unbalanced and inflexible.

This type of repetitive stride length can also have an impact on the joints. If there is a constant pounding of the ground at one range, the joints will be constantly impacted over a small area, creating a high impact in a reduced area, for which they are not designed.,

Also, with this type of fast, tiring, highly stimulatory exercise, it can be difficult for a dog to settle and recover through not being able to relax and sleep – 'rest and repair'! This can then lead to a situation where the human or guardian can be misled into believing that their dog needs more exercise (see turbo puppy, page 111; calming exercises, page 151).

If this is their exercise before everyone in the household leaves for work and the dog is then left at home, the guardians may not know that the dog is in a heightened state and cannot settle. They may be restless for a long period of time, never allowing their body to recover.

This is the *most* unsuitable activity for a puppy or adolescent.

MANAGING AN OVERHEATED DOG

If your dog becomes distressed with heat and panting is not cooling them, putting cold water over a hot dog's body can be highly detrimental. By doing this you are preventing heat from escaping through the blood vessels in the skin, which is a major way a dog cools down (in addition to panting and sweating slightly through its feet). Rather than the vessels being dilated, allowing heat to be lost, the vessels would then constrict due to the cold water. The dog is then hugely disabled from naturally removing heat from its 'core' body or internal organs. This is highly dangerous and can cause hyperthermia, and even death.

Immediately take them to a place where the air is cooler, so they can inhale cool air into the centre of their body, then place something cool over their belly and legs.

'SIT!' WHAT ARE THE PHYSICAL EFFECTS OF REPETITIVE SITTING?

Training a dog to 'sit' has manifested into an everyday interaction between dogs and their guardians and is taught in every puppy and dog obedience class across the world. The puppy's guardians are encouraged by professionals to ask their dog to sit in almost any context to manage their puppy's behaviour. There is no doubt that training behaviour is important to help our puppies succeed in our human world, but we must also consider how a puppy's body is built and what it is designed to tolerate and not tolerate.

A 'sit' seems like such a simple and natural action, and it appears to be an action a puppy adopts when they need 'thinking' time. This 'thinking sit' has the physical appearance of being more postured than other types of 'sit' (Figure 7.14).

So why do puppies sit if it could potentially cause damage? The question is not about the action of sit and stand, but again it is the unnatural repetitiveness of the action being requested by traditional training methods (Figure 7.15).

There is no doubt that dogs and puppies are anatomically designed to sit. However, if puppies are habitually and unnaturally asked to sit, the loading on their

203

Figure 7.14 Puppies often appear to adopt a more postured sit when they need to think, or a 'thinking sit'.

Figure 7.15 Puppies do choose to sit naturally, often when they are thinking.

body could have an accumulative effect as they progress into adulthood. Especially if we add the other potential environmental factors that can also cause damage such as running on slippery floors and repeatedly jumping on and off a sofa.

To put a human perspective on this, most primary school-age children are happy to sit on a floor cross-legged. Fewer adults would be as comfortable adopting such a pose for any length of time. So as adults we can choose not to sit in this position.

However, our dogs are not given this choice because of the conditioning to go into a sit from puppyhood. We teach our dogs to routinely sit throughout the day, before being fed, sitting at a kerbside, before receiving a treat, and so on. As a dog ages, this conditioned response continues because they associate a sit with all these behaviours so they continue to offer it despite the discomfort they may feel.

To understand the impacts of a sit, let us look at the biomechanical action of a stand to sit, and sit to stand in an adult dog with no known underlying issues.

THE BIOMECHANICS OF STAND TO SIT

The stars indicate where the main forces are being driven, static, rotational, and kinetic (Table 7.2).

The pictures were intentionally taken with the dog's head elevated, to replicate the posture when responding to instruction.

 key area for loading
★ major area of fixed loading

Table 7.2 The Biomechanics of Stand to Sit	
	When dogs sit, they must hold their weight against gravity during the process. A large percentage of that load and the cantilever force goes through their elbow. They tend to dominate one elbow rather than equally spread the load. There is also a load through the neck and shoulders using eccentric contraction (muscles having to hold a load and allow a controlled muscle fibre release). Excessive challenge can be damaging to muscle fibres.
	The elbow is really taking the load to ease the settling into a seated position, which will release the loading on the knee. The knee or stifle is now taking up some of the stress and load, as is the carpus or wrists, of both the front legs.
	The neck now takes up some of the load, but the elbow continues to brace the body as the pelvic region meets the ground in a sit.

(Continued) **205**

Table 7.2 (Continued) The Biomechanics of Stand to Sit

(d) At this point, the elbow is still bearing weight. Even though the dog looks like she has sat, she has not fully settled her bottom on the ground.

If we now look at this final stage of the sit more closely, you'll see that the dog goes almost from a hover (5), then settles onto the 'seat bone' of the pelvis (6). When the bottom finally settles onto the ground, there is a large forward thrust from the tibia towards and into the stifle joint, which also impacts on the hip, until the weight is fully settled onto the ground.

From the images, you can see the main forces that are being driven and the areas of the body that are being loaded when a dog sits, with a large percentage of the load going through the leading elbow. Now let us look at the biomechanics of sit to stand.

THE BIOMECHANICS OF SIT TO STAND

See Table 7.3.

 key area for loading

★ major area of fixed loading

Table 7.3   The Biomechanics of Sit to Stand

	The dominant elbow takes up the load. Other areas are preparing to lift the body off the ground.
	The body here must support itself fully against gravity and swing itself forwards into a stand. The elbow, the wrist, or stifle, and hip are all under massive load and kinetic force.
	The pelvic region and lower back are supporting the body as the forequarters start to take more of the load and the forwards trajectory into the initial stride, which will be taken up by the dominant elbow, until the body is standing on all four legs and can balance the weight and load.

We can see that once again the dominant elbow takes a large percentage of the load during this action and as with most physical activities, it is the repetitive nature of the action that can be detrimental. With puppies in particular, their skeleton is developing, and their skeletal structure grows quicker than their associated musculature soft tissue attachments. Therefore, the puppy with proportionally less developed muscles, especially over the hindquarters, will find it difficult to push themselves up from a sit. They will have to adjust their hind limb position to aid and create the required drive and force (they do this by drawing the hind legs and feet closer together to combine strength). However, much of the force and drive will be also driven through the forequarters or forelimbs.

To summarise, it is responsible for us, as dog guardians, to be aware of what we are training our puppies to do, as well as awareness of their environment and activity levels, which may be having a negative impact on their physicality that will affect them in later life.

JUMPING UP REPETITIVELY

Puppies will generally jump up: it is a great way of gaining our attention and it is difficult to avoid with even the smallest puppy. However, it's generally quite unsociable for your puppy, and then your mature dog, to jump up at other people. More importantly, it is not good for the skeleton and muscular connections (Figure 7.16).

Jumping up creates a load down the puppy's vertebrae (see vertebrae, page 16), compressing those vertebrae. It also puts excessive and unnatural repetitive load onto and through their hip joints. Maintaining the position overuses their hamstrings (dotted line), which ultimately shorten through the fibres, and can then adjust the positioning of the pelvis.

Jumping down creates a massive force going into the puppy's shoulders (page 30) and lower neck (see page 19).

PLAYING EXCESSIVELY WITH LARGER DOGS – ESPECIALLY ADULTS

It is so important for a puppy to interact with all dogs, young and old alike. However, be aware that older mature dogs will be heavier, stronger, and more worldly wise. Most of the time these social interactions are to be encouraged. However, be aware of not letting your puppy run with bigger dogs for **extended, repeated periods of time**. Running with dogs with longer legs will encourage your puppy to overextend to catch up. When they are in their developmental stages, this can have an impact on their joint development (Figure 7.17).

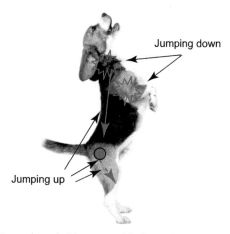

Jumping down

Jumping up

Figure 7.16 Jumping up at people and objects repetitively can have ongoing concussional effects throughout the whole body.

Figure 7.17 Dogs come in all different sizes, try to limit the amount of play with dogs much bigger and stronger than the puppy or young dog.

KEY POINTS

- To protect our puppy's physical health, it is equally important to consider the things *not* to do, as well as what to do!
- Repetitive strains are insidiously destructive to a dog's physical (and mental) body.
- Dogs just want to be with you, and will not actively complain, so you *must* be their voice.

8 *Equipment*

It is not the equipment itself, but how it is used. Consider the effects any equipment has on your puppy when 'used'!

There are so many adverts, 'professional' opinions, and options regarding the equipment your puppy or dog needs.

There is a great deal of good equipment that will assist you in keeping your puppy and mature dog healthy and, most importantly, safe. However, like all equipment, it is not the tool itself that's important, but how it is used in *all* situations.

Equipment should never be used to replace good education or training (for both you and your puppy). For example, collars, harnesses, or head harnesses that stop dogs from pulling. As a responsible individual, you should find out how to positively encourage your dog to stop pulling, rather than use subversive means.

If, however, that same piece of equipment is going to stop you both from having an accident, then you *must* have the proper training to understand how and when to use it appropriately. You must also understand how it affects your puppy's anatomy.

RESTRAINING EQUIPMENT: COLLAR, HARNESS, HEAD HARNESS

One of the most important pieces of equipment you will need is for taking your puppy outside when you need to keep them safe. We are so conditioned to use a collar that perhaps we don't even give the process of wearing a collar the correct consideration.

As with all equipment we use with our puppy and mature dog, we must consider the effects on their anatomy; that is, whether it is damaging or restricting their body. So many of these branded methods of restraint do not consider this vital fact, creating potential injury and 'pinch points' within the puppy. These can create injury, both seen and potentially unseen, within the internal anatomy and physiology of your dog.

It is about being aware of the potential issues that can occur with different mechanisms of restraint but also providing safety for you and your dog.

Ancillary considerations: Where and how the lead attaches are important considerations.

Harness and leash clasp: Ensure that both the leash and harness clasp do not flap and hit the dog's back. Also, make sure the leash attachment is not too heavy, so it will not hit the puppy's back. After some time, these two mechanisms could create damage over the structures of the vertebrae.

Identity disc: Likewise, ensure that the identity disc is least likely to create wearisome tapping on your puppy or dog's body.

211

DOI: 10.1201/9781003268789-8

Remove the collar/harness at night: If it is safe to do so, in the home, give your puppy time off from wearing the collar or harness. Clothing can irritate our skin after a period of time, and collars or harnesses will possibly have the same effect on them too. It is also worth remembering that collars can get caught in furniture, or in other dogs' teeth when they are playing.

Consider the weight and type of leash attachment you have to your dog. They can be very heavy and repetitively rebound as the dog walks and runs, creating at best an annoyance and worst a point of injury or pain (Figure 8.1). This can be relevant with a collar or harness, especially as harness attachments are generally over the top of the dog's vertebrae.

THE LEASH

There are so many leashes available and there are some wonderful studies and papers on the optimum length of leads to create the safest connection between you and your puppy and dog, whilst allowing them freedom of movement. Ten metres is optimum when there is good open space, 5 m for quieter areas, and a minimum of 3 m when in a busy environment.

THE RETRACTABLE LEASH

These are really popular as it is felt that they can offer the dog or puppy freedom of movement but also safety by being able to retract the leash.

Figure 8.1 The leash connection and identity tag are the attachments that harnesses and collars support. These can be a repetitive irritation to a dog.

Unfortunately, often the leash can present the worst of both situations. A dog can suddenly run, and they are too quick for their human to put on the brakes, resulting in a total lack of control. The dog could be running towards something and manage to out-spin the leash and the handler, resulting in the dog running too fast towards danger or, worse still, running into the middle of danger, such as a road (*something I have witnessed first-hand*) (Figure 8.2).

Another situation is when a dog suddenly runs forwards and the handle is wrenched out of the handler's hands. The noise of the mechanism retracting and bouncing behind them scares the dog so much that they continue their run out of fright. The same scenario can occur with the leash mechanism hitting the dog or puppy.

It is also possible for a puppy to suddenly run when their guardian is not ready, with an opportunity to gather speed due to the length of the retractable leash. When it suddenly reaches the end, the dog is catapulted into a sudden stop that could cause a whiplash injury.

Many of these leashes are made up of thin cord, which can create a burn or abrasion over the skin if the dog becomes wrapped up. This cord can also break under tension after repeated use.

It could be better to get a long leash that you can clip to make shorter when you need to keep your puppy safe and close but allows an unclipping that can extend when your puppy needs a little more exploration space.

Table 8.1 gives information on the different types of restraints, with an objective view of their application and rationale. **Remember that human and canine anatomy is almost identical.**

Figure 8.2 A retractable lead can reduce control of your puppy, especially in an emergency situation.

Table 8.1 Equipment Type

Equipment type	Wearing the equipment	Pulling on leash	Potential damage from misuse	Safety advantages
1. Collar	The collar is fairly benign if there is no pressure going through to the neck unless it is too tight. How comfortable would a human be wearing a neck brace for control?	**Localised damage:** The consistent compaction and compression of all the critical neck structures. Also, leaning into the lead consistently can compress the vertebrae by the drive from their hindlimbs through to the restriction by resistance by holding the leash. **Global damage:** The drive being asserted through their hindlimbs will create an anatomical alteration to the joint integrity through their hind and forelimbs. This will create unbalanced musculature that will have an impact on these joints even when they are not pulling on the leash. **Lurch or leaping forwards:** If a puppy or dog lurches forwards with a collar, this can create an injury, or even a whiplash, which could go totally unnoticed until it creates secondary issues. *Neck compression injuries could impact not just the central nervous system but also the cranial nerves, especially the vagus nerve which has a huge regulatory impact on all types of behaviour and stress moderation.*	The collar enables a stranglehold on the puppy or dog, so if they are required to move, or be lifted, the collar might be incorrectly used.	The collar is the least likely restraint for a puppy or dog to escape from.

(Continued)

Table 8.1 (Continued) Equipment Type

Equipment type	Wearing the equipment	Pulling on leash	Potential damage from misuse	Safety advantages
2. Slip lead	The slip lead is completely benign when being worn without any pressure. However, it is effectively the same mechanism as a noose (Figure 8.3). Figure shows a common design of a slip lead, they are very quick and easy to put over the dog's neck but act with the same mechanism as a noose.	**Localised damage:** The consistent compaction and compression of all the critical neck structures. Also, leaning into the lead consistently can compress the vertebrae by the drive from their hindlimbs through to the restriction by resistance by holding the lead. **Global damage:** The drive being asserted through their hindlimbs will create an anatomical alteration to the joint integrity through their hind and forelimbs. This will create unbalanced musculature that will have an impact on these joints even when they are not pulling on the leash. **Lurch or leaping forwards:** If a puppy or dog lurches forwards with a collar, this can create an injury, or even a whiplash, which could go totally unnoticed until it creates secondary issues. *Neck compression injuries could impact not just the central nervous system but also the cranial nerves, especially the vagus nerve which has a huge regulatory impact on all types of behaviour and stress moderation.*	Due to its mechanism, the potential for strangulation to different degrees has a high risk.	The slip lead is the least likely restraint for a puppy or dog to escape from.

(Continued)

Table 8.1 (Continued) Equipment Type

Equipment type	Wearing the equipment	Pulling on leash	Potential damage from misuse	Safety advantages
3. Head harness	This can be uncomfortable for dogs to wear. Having a strap over their nose and just under their eyes can be very irritating and can create chafing. Some of these halters can be misused very easily and need professional advice on their use. The material on some of these halters is insensitive material for areas of sensitivity over the dog's face and head. These should only be used in extreme situations. Figure shows one of many types of head harness/halti equipment. The sliding mechanism shortens around structures over the face, often with harsh fabric.	There are many different makes and models of these, and they all have slightly different applications – some are more severe than others. **Localised damage:** The potential for using the leverage force through the neck could create many different neuro-muscular problems. They can also move from a position and seat themselves over or just under the dog's eyes. They can also promote localised neck damage. **Global damage:** The leverage over the neck can disrupt the biomechanical chain, affecting the whole movement pattern. The drive being asserted through their hindlimbs will create an anatomical alteration to the joint integrity through their hind and forelimbs. This will create unbalanced musculature that will have an impact on these joints even when they are not pulling on the leash. **Lurching or leaping forwards:** Potentially cause a whiplash injury or injury to the atlanto-occipital joint (where the skull joins the cervical vertebrae)	This equipment has been developed for maximum levels of leverage through the head and neck. Misuse can potentially cause neck whiplash injuries. If it is incorrectly fitted, it can chafe by being too close to the eyes.	They provide control for large or very strong dogs through restriction and leverage. For highly nervous or reactive dogs, it can be an aid to draw the dogs' eyes back to the human for security and reassurance that can prevent escalation through any confrontation.

(Continued)

Table 8.1 (Continued) Equipment Type

Equipment type	Wearing the equipment	Pulling on leash	Potential damage from misuse	Safety advantages
4. Harness with chest restraint **Not all harnesses are the same!**	These harnesses are restrictive from the moment they are put on. They restrict the free movement of the shoulders, as well as potentially the movement through the thoracic region of the vertebrae. These harnesses also tend to be of an inflexible design that is the equivalent of wearing an incorrectly designed and positioned saddle. Figures show another style of chest restraining harness; and a chest restraining harness, supporting a saddle and handle.	**Localised damage:** This is a highly restricting piece of equipment, yet interestingly, it is designed to look like a serious working dog's harness. This has been designed with the best of intentions but not taking the canine anatomy into account. With the strap across the shoulders and chest, it is the equivalent of us trying to walk or run with our arms strapped down at the shoulders. Like us, without free movement, the biomechanical chain is going to be interrupted and this will create adaptation and compensation through the body. **Global damage:** Not having free movement through the dog's forelimbs will have an impact on the hindquarters, potentially reducing the kinetic chain and reducing power and fitness. This impedes correct and balanced muscle development. It also reduces lateral flexion through the thoracic region, which will have an impact on a dog's biomechanics, and cross-lateral strength (see page 143). **Lurching or leaping forwards:** This will not have such a negative impact due to the wide area of restraint, but it enables the dog to assert great force into a forwards trajectory, potentially reducing significantly any control in a potentially dangerous situation.	If the dog spends a long time in this harness, or spends considerable time running or jumping up or down in a work or sport situation, it could have long-term effects on muscle and joint health.	They provide a reasonably safe and easy to grab handle for removing your puppy away from danger.

(Continued)

218

Table 8.1 (Continued) Equipment Type

Equipment type	Wearing the equipment	Pulling on leash	Potential damage from misuse	Safety advantages
5. Harnesses with multiple adjustments	Harnesses should always fit the puppy or dog over specific regions. Some that are adjustable can be good when they provide free non-chafing movement. Some harnesses have so many clips for alternation that these can cause localised soreness. The harness should fit so that the chest connection of the harness still sits over the breast or chest bone when under pressure. The arm straps should allow free movement of the forelimb	**Localised damage:** A harness without a long enough 'back strap' will consequently not finish far enough down the dog's back and can cut in under their 'armpits'. This is a very sensitive neurological region. Any impingement whether it is neural or muscular will be continually aggravating, with chafing as well as damaging. If the front joint over the chest is not fitted correctly, when pressure is applied on the leash, the harness can in effect compress all the vital structures under the dog's neck such as the trachea, oesophagus, glands, muscles, and blood vessels. The multiple clasps can also create irritation. The fabric can give the impression of being cushioned by padding over the strapping, however, when the harness is pulled, this padding can be almost ineffective, creating a very targeted, narrow pressure over the dog's body. **Global damage:** Restriction of the shoulder movement will have a negative impact on the drive of the dog by causing a break in the biomechanical chain. The drive being asserted through their hindlimbs will create an anatomical alteration to the joint integrity through their hind and forelimbs. This will create unbalanced musculature that will have an impact on these joints even when they are not pulling on the leash. A constant chafing could have an impact on the dog's attitude and behaviour during the times of being worn and could leave them with residual tenderness. **Lurching or leaping forwards:** If the chest connection is incorrect, this could severely dig into the ventral (underneath) structures of the dog's neck.	If it is badly fitted and left on the dog for extended periods, it can cause physical and potentially psychological issues, caused by continual irritating chafing.	This is a relatively safe harness that mainly prevents dogs escaping from the harness and allows for control.

(Continued)

Table 8.1 (Continued) Equipment Type

Equipment type	Wearing the equipment	Pulling on leash	Potential damage from misuse	Safety advantages
6. Sliding harness	When the dog is still there is little or no pressure on the body	This harness is often made of cord and is on a slider, so when the dog pulls, the cords restrict around their whole forelimb anatomy, compressing both soft tissue of the shoulder and the lower neck.	Constant repetitive pressure over this area could have a major impact on the musculature as well as potential neural damage under the elbow if pulled very tightly.	They are unlikely to slip out of it.
7. Fitted harness	A well-fitted harness made of a good quality soft fabric should sit on the dog's body, allowing for free movement without causing irritation. It should fit over the chest, allowing free shoulder activity, and if pressure is asserted, it is being placed over the sternum, not the trachea. Figures shows the anatomical areas on a dog that are important to recognise when fitting a harness.	**Localised damage:** If used correctly there should be minimal damage, due to the harness fitting the dog's body, rather than the dog's body being made to fit into the harness. the fabric should be soft and wide enough not to cause any pinching or digging into the dog's body. **Global damage:** If correctly fitted and used, there should be minimal damage. **Lurching or leaping forwards:** If correctly fitted, it should not damage the dog. However, if pulling is persistent by the dog, the same issues could occur as other equipment. The drive being asserted through their hindlimbs will create an anatomical alteration to the joint integrity through their hind and forelimbs. This will create unbalanced musculature that will have an impact on these joints even when they are not pulling on the leash.	Correctly fitted, this type of harness should not cause damage.	If the harness is fitted correctly, they are unlikely to slip out of it and you can hold the dog close to you via the back strap, without damaging the vital anatomical structures, such as in the neck.

Figure 8.3 Other types of slip lead can be used when showing. These leads are placed so that the drawstring is high up on the neck to facilitate a high head action.

WHAT HAPPENS TO THE PUPPY'S BODY WHEN USING THE EQUIPMENT?

When choosing the correct equipment for your puppy (and your mature dog), an important consideration is the effect the use of this piece of equipment is going to have on your puppy's physicality when it is put under load or strain.

Observe what is going on in your puppy's body:

- When restraining your puppy.
- When walking/running/standing up from a sitting or lying down position.
- Where it pinches or rubs in everyday situations.
- Where it pinches or rubs in an active situation.

Equipment fits very differently when the puppy is active.

THE COLLAR AND THE SLIP LEAD

A collar will obviously restrict the neck and all the vital structures arising from and running through the anatomical region (Figures 8.4 and 8.5).

When using a collar, we must remember the anatomy that lies beneath the equipment; the neck is the conduit for 'life'. If the neck is constricted by using a slip lead, choke collar, or even a fixed buckled collar, it can cause **strangulation**, just

Figure 8.4 What lies under the neck and where the collar sits on the neck of the puppy.

Figure 8.5 The strangulating effects a collar can inflict if there is resistance.

as it would if the same apparatus were attached to a human neck. Strangulation through the neck is the constriction of structures that provide the literal 'life blood' for the body.

The neck facilitates these structures and mechanisms, and a tight collar will restrict them.

1. A passageway for oxygen to the lungs and to expel toxic carbon dioxide (trachea).
2. A passageway for food to enter the stomach provide nutrition for the body (oesophagus).
3. The whole neural supply from the brain, including the brainstem that controls breathing and the heartbeat and is responsible for life (central nervous system).

221

4. The peripheral neural supply for movement and sensory understanding (peripheral nervous system).
5. The blood supply to and from the brain, supplying the brain with nutrition to maintain its continual uninterrupted performance (carotid blood supply).
6. The blood supply for the sensory organs of the face – eyes, ear, nose, and mouth.
7. Vital glands that feed information to and from the whole body that in turn give direction for immunity and balance to other organs and glands within the body (thyroid and lymph).
8. Vital muscle and fascial connections are critical for the activation and operation of full-body movement.
9. Vertebral structures that support the nervous system, and if integrity is lost, create high levels of pain (cervical vertebrae).
10. Bony structures and soft tissue support the function of the tongue (hyoid apparatus) (see page 13).

THE HALTER/HEAD HARNESS

These types of harnesses are normally chosen by people that are having a problem holding their dogs whilst on a lead. A dog not under control is potentially a huge risk to themselves as well as other people and dogs.

A dog pulling consistently on a lead can create a massive inappropriate load with strains on its body, so it is vital from all perspectives that we can walk *with* our dogs safely, happily, and with as little harm as possible.

If this is the restraint that you have been recommended by a qualified canine professional trainer, for your specific situation, please ensure you regard the expert tuition and you are fully aware of the implications on your dog involving its use.

The head harness comes in various designs with different applications for use. Some create a much harsher effect than others, which can be relatively gentle, *if correctly used and professionally trained.*

This applies pressure over one of the dog's most vulnerable areas, between the skull and the cervical vertebrae. The area of compression is the joint between the two and has limited protection. The sliding, compressing action of this type of headcollar can cause serious issues over this vulnerable region that will not be immediately apparent (Figures 8.6–8.8).

Figure 8.6 Mainly due to the leverage through the nose from these pieces of equipment, they can potentially create and cause an extreme neck injury if they lunge forwards when being securely restrained.

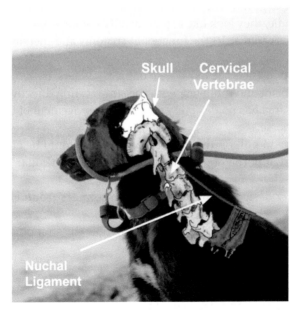

Figure 8.7 Where the equipment fits in relation to some of the cervical or neck anatomy.

Figure 8.8 A picture of the placement over a head harness style restrainer.

THE HARNESS – NOT ALL HARNESSES ARE THE SAME!

CHEST RESTRAINT HARNESSES

There is great popularity with these chest retraining harnesses, not just with pet dogs but working dogs too. They look 'workman-like', and with the saddle-type back and the handle, they look like a good quality working harness. However, these styles of harnesses can restrict the functional anatomy and therefore the movement of the puppy and dog.

The chest strap restricts the movement of the shoulders, making the extension of the forelimbs difficult and inhibited, with the strap resting directly on the shoulder joint. The saddle area has no flexibility and restricts movement of the scapulae, akin to a saddle on a horse.

If the dog were to pull on the lead, the wide strap could be elevated and compress the throat of the dog, potentially the same as a collar would in the same situation.

There are other designs of harness that also connect around the front (Figure 8.9).

These pictures give a visual demonstration of how the strapping over the chest of the dog can restrict movement.

ILL-FITTING HARNESSES

Dogs come in all shapes and sizes, so taking a little time to ensure the correct fit is vital for your dog's safety and well-being, and also their overall perception of a walk. It should be a fun, comfortable experience and if their equipment creates discomfort, their 'fun' time will create an immediate stress. Also, if it hurts when they pull to sniff or meet another dog, it could set up a bad association with that event.

There just isn't a 'one size' for all harnesses, but the one that would be best will have certain criteria that will enhance the 'fit' on your dog's body.

Figure 8.9 Pictures showing how having a strap over a dog's chest can also hinder them from getting up from a sit or a down.

The best harness is the one that:

- Fits *your* dog, and not your dog made to fit the harness.
- Has a long enough top back strap to allow free movement of the forelimbs, when still or active.
- A 'y' style front that sits on or just below the sternum or (manubrium).
- Doesn't chafe or rub when they move (Figure 8.10).
- The dog cannot escape from.
- Allows full free movement of forelimbs.
- Does not constrict the trachea and lower (or upper) neck structures when pressure is asserted through the leash (Figure 8.11).
- Has fabric that is strong enough but equally soft and wide enough to be comfortable on the body.
- The leash fixing point doesn't flap over the back (vertebrae).
- The leash clasp doesn't flap and continually hit the back of the dog (vertebrae).
- When pressure is asserted on the leash clasp, the 'y' connection doesn't compress the throat.

This harness fits the dog's body and is also safe and comfortable to wear.
See the figure of the fitted harness in Table 8.1.

Figure 8.10 One of the most common fitting mistakes is for the top back strap to be too short.

Figure 8.11 A harness that may fit when the puppy is standing can impact the dog's anatomy when there is pressure.

KEY POINTS

- We do not *mean* harm to our puppies and dogs, it is just that some equipment is used without consideration, which could be uncomfortable or even harmful.
- It is vital that if we are having problems with our puppy's walking safely on the leash, we should gain proper expert advice.
- Each dog is a different shape, we should get equipment to fit their body.
- When fitting, consider how the equipment will fit your dog's body when active and/or put under strain.

9   *Puppy massage*

When a puppy leaves its mother and siblings, they immediately leave the comfort, security, and stimulation that the environment provides. When your puppy arrives at their new home, it would be great if we could continue to provide this type of contact and stimulus to maintain this close and secure relationship.

Massage has both physical, physiological, and psychological advantages. Psychological and/or emotional can be from the soothing contact that your puppy could be missing (Figure 9.1). Physiological because it influences good blood delivery and waste disposal.

The arterial blood that is pumped from the heart distributes fresh oxygen and nutrition-rich blood to all the tissues of the body. This is a highly effective mechanism for delivering what the body requires as far as oxygen, nutrition, including all the vitamins and minerals, and hormones.

The arterial system is highly effective. However, the venous (vein) system that carries the blood back from all over the body is the system that transports all the metabolic waste from the cells. The veins do not have the advantage of a blood pump that is the heart. The blood in the veins relies on being pumped back through the body to get the toxins away from the body, predominantly by physical activity and alternating the body's internal pressure changes. But mainly by the movement of the body. Massage, especially effleurage, can really aid this process.

Another advantage of having positive direct 'hands on' contact with your puppy is that you become more aware of how your puppy 'feels'. You begin to recognise the normal temperature fluctuations of their body and the 'feel' of his or her muscles and joints. This will allow you to have an idea of what is normal. If there is a problem involving a potential injury or other such health issues, this knowledge gained by you could be invaluable should it need further investigation by your vet.

The benefits of massage from your puppy's perspective:

- A puppy becomes used to being touched and handled (important for initial routine vet visits).
- Further helps to create a bond of trust.
- A greater bonding and connection between puppy and their human.
- The enhanced feeling of security – they came from a very tactile environment and now could be on their own.
- It can aid to increase a puppy's deeper breathing, therefore de-stressing the puppy naturally.
- Enabling a puppy to take deeper breaths will help its brain and body to function better.

DOI: 10.1201/9781003268789-9

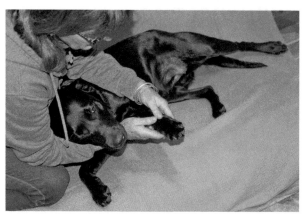

Figure 9.1 Massage can be really bonding for you and your puppy.

The benefits of massaging your puppy from the human perspective:

- Puppy–human bonding and attachment. This is a tangible feeling and emotion.
- A positive feeling that helps the puppy to trust its environment.
- Having a literal hands-on knowledge of a puppy's physical condition.
- Start to recognise areas of issue, such as inappropriate constant topical heat (felt through the skin but could be radiating from deeper within their anatomy) over a particular part of the puppy's body that may indicate an injury or repetitive strain.
- Get to know what is normal on the puppy's body, so the identification of lumps and swellings can be quickly identified and can be treated quicker.

Health benefits for your puppy:

- Massage can encourage good blood flow that will help muscle function, and so assist good functional movement.
- It gently encourages venous return, or the veins to remove blood with metabolic toxins.
- Encouraging the flow of venous blood allows and encourages arterial blood delivery. This helps with growth and cellular development, mineralisation of joints, and hormone delivery. This is really important for growth, especially in the larger breeds (or breed types).
- Have hands-on information about your puppy's changes, and to some extent, information about their physical health.
- It encourages lymphatic stimulation that supports their immune system.
- It can be calming for the puppy.
- The release of bonding hormones can be stimulated when positive visual and physical contact is used

- Gentle positive touch can be especially beneficial if the puppy is particularly nervous or had a bad start in life
- Massage can assist your puppy to be aware and help them to 'connect' their body.
- It can help puppies to relax, stimulating the parasympathetic nervous system, which is the 'rest and digestive' zone, aiding digestion (see page 68).

HOW TO START

This will be a slow process, being a new experience for your puppy and possibly for you too! Do not set your aims too high. A short two-minute positive session will be a great start.

Getting your puppy used to you putting or laying your hands over his body is a very important socialisation lesson. It will help get them used to being touched if they need to be examined by yourself or a vet.

STARTING TO MASSAGE YOUR PUPPY

A great way to start is to have a specific piece of bedding or something that will indicate what you are going to do. This bedding will then be a constant when you are going to do some massage.

By adopting this mechanism, it will communicate to the puppy what you are going to do, and it should have a good connotation. Eventually, it will enable the puppy to choose if they stay or go. Allowing your puppy their autonomy so early is very powerful for them, and it will help them learn to trust you and give them more confidence. It is incredible how quickly a dog and/or puppy will work out and learn the process. This choice-led treatment has always been used by Galen Myotherapy for all their treatments, it is an incredibly successful and scientifically supported approach. It is called Positive PACT®.

It is always recommended to massage on the floor, but you can start with your puppy on the bed or bedding, sitting on your lap, with you sitting on the floor. This is to allow the puppy to safely leave if they want to. You are looking for just a few seconds of massage at a time to start with – patience is required!

When they leave, just sit, and invite them back very calmly; in time you will become a safe place for them for both treatment and security.

Start with very short timescales (max five minutes) and you can build up to approximately 30–40 minutes per week or a few minutes a day by the time your puppy is 12 months old.

The more you try this, the more you will discover about your puppy.

You can move gently over their body, working over their head, ears, and feet, all areas that at some time they will need some form of handling or investigation. This will offer them a really positive experience.

The best times to massage:

- Do not massage two hours either side of feeding time.
- If possible, it would be best just before they sleep or when if they are quiet.
- Don't try when there is too much other stimulation going on in the room.
- If possible, work with your puppy's timing – the evening is potentially a good time.
- If you are feeling stressed it is best not to start. The feelings will be reflected through the treatment.

The techniques suggested:

- Passive touch
- Effleurage
- Skin rolling
- Paw work
- Face and ear work

Effleurage is a stroking technique that is done with the flat of your hand, with your fingers all close together and your thumb lying next to your fingers to form a flat surface (Figures 9.2–9.5). Be sure that an equal pressure is being asserted through your whole hand and fingers. Conduct the technique by using long flowing 'strokes' using either one hand or two, with one following the other (Figure 9.6).

Always go in the direction of the lie of coat (as can be typically seen in a Labrador's coat) (Figure 9.7).

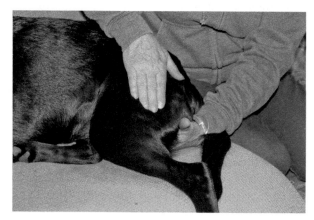

Figure 9.2 The whole hand using effleurage over the hindlimb – placing a hand under the stifle to protect the joints.

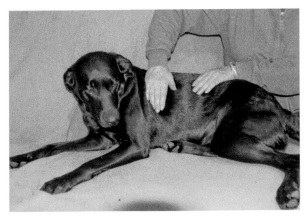

Figure 9.3 Effleurage, a stroking technique, using the whole surface of your hand, to create an even pressure. Always keep the other hand gently in contact with the puppy's body to ensure constant contact.

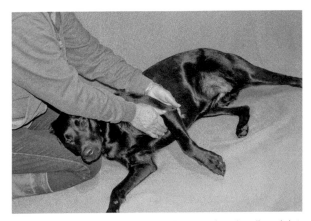

Figure 9.4 Performing effleurage under the chest and supporting the elbow joint.

Figure 9.5 When your puppy or dog is small, you can apply effleurage using just two fingers instead of a whole hand, being extremely careful not to assert too much pressure over the smaller area of contact.

Figure 9.6 Whilst the head is cradled, the thumbs can then gently circle over the cheek muscles.

Figure 9.7 Indicating the directional lines where to apply massage (the dotted lines are under the body), the red marks indicate the joints to support. It is better if the puppy is lying down, but you can still apply massage with them standing.

FACE MASSAGE

Forehead gliding – using the flat of your thumbs, you can use both simultaneously, as this allows you to cradle their head in your hands. Gently glide the flat or pad of your thumbs up and over your puppy's forehead. Puppies can hold tension here, so it can be extremely relaxing for them.

    Cheek circling – like their forehead, their cheeks can also hold tension. The cheek muscle, or masseter, is a circular shaped muscle, so replicating that with the massage stroke can have a very positive effect. This too just needs gentle application using the pad of one or two of your fingers.

Ear gliding – a good calming technique is ear gliding, which would come under the title effleurage. Gently take your puppy's ear between your forefinger and thumb and starting at the point nearest their head, draw your finger and thumb down the ear flap. You can repeat this action several times. This is a very calming technique and can be used if your puppy is anxious (Figures 9.8 and 9.9).

Points to remember:

- When you start, gently lay your hands on your puppy's body for a couple of seconds to introduce them to the touch and prepare them.
- Apply effleurage along the natural lie of the coat (see Figure 9.7) – this is very important.

Figure 9.8 The direction and areas of the face to target for a puppy face massage.

Figure 9.9 Gently cradling the head and allowing the thumbs to gently perform effleurage over the top of the head and following through into the ear glide.

The application of canine massage differs from human massage – it is not required to work towards the heart, it is better to work down the functional fascial lines (see fascia, page 94).

- Use long, rhythmic, gentle strokes.
- Ensure the pressure used does not exceed that of a normal stroke. (If you apply excessive pressure, you could damage your puppy's muscles and stress its joints.)
- When you work over joints, be sure to support them with your non-working hand.
- Always ensure that you have one hand in constant contact, so if you are using one hand to effleurage, let the other hand gently rest or support your puppy.
- Do not work over the top of his vertebrae or spinal column. Work either side.
- When you finish, as you did for the start, gently rest your hands over the area that you were working for a couple of seconds before you lift your hands off.

PASSIVE TOUCH

Passive touch is an active technique even though we appear to be doing nothing! Place your hand over an area of your puppy that they are comfortable with. Place your slightly 'cupped' hand over the area for a few seconds.

It not only has a comforting effect, but it also subtly warms the area allowing for a slight vascular change (blood flow change). This calms the nerve ends that can help with an area of slight tension. Areas such as elbows, shoulders, backs, and hips are particularly good areas for this very gentle technique (Figure 9.10).

Figure 9.10 Passive touch, a slightly cupped hand gently placed on the puppy's body.

SKIN ROLLING

This is possibly one of the most versatile techniques that will not only help to relax your puppy but will keep their skin flexible and healthy. It can also help to maintain good superficial fascial flexibility.

- Gently lift the skin with your fingers and thumb in one or both hands.
- Walk fingers, forming a wave of skin, down your puppy's shoulders towards his elbow, with your thumb following, forming a gentle and continuous ripple of skin gradually moving from his back down to his elbow.
- Avoid pinching the skin as you lift.
- Do not perform if the skin will not lift with ease.

Continue the action on areas on the top of your puppy's neck and down their back. If the skin is tight do not attempt to lift as this will be very uncomfortable. Just use this where the skin is loose and flexible (Figure 9.11).

ADDITIONAL TECHNIQUES

To help neural development and spatial awareness in your puppy, you can also incorporate this specific massage over your puppy's paws (it can also be used on older dogs too!). It will also get them used to having their feet handled, as this can become a problem later on in their life.

Paws – perform each stroke several times on all paws, ideally with your puppy lying on the floor.

1. **The pastern slide (top)**

    Stroke the front of the paw a few times then lift the paw and cup it in your hands for a few seconds. Then gradually with alternate thumbs stroke the top of the paw, starting from the wrist or carpal joint and going all the way down the pastern to the claws (Figure 9.12).

Figure 9.11 Skin rolling, gently lifting the skin and rolling it across the dog's body.

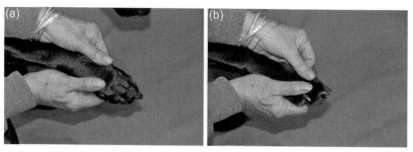

Figure 9.12 The pastern slide.

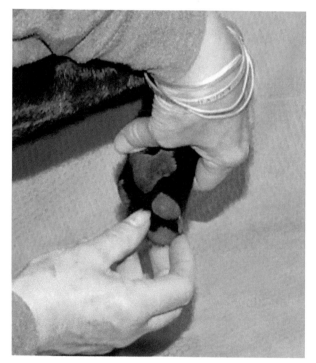

Figure 9.13 The pad squeeze.

2. **The pad squeeze**

   Cup the paw in both hands for a few seconds. Then very gently **press the pads** with flat fingertips – all over the pads. One hand supports the paw and the other hand gently applies the technique (Figure 9.13).

AFTER MASSAGE

When you have finished any form of massage, gently rest your hand over the last area you were working on. This will prepare your puppy for the 'end'. Otherwise, if they have been relaxed and enjoying the attention and you suddenly leave, it will create a minor stress. They will be unaware that the session has come to an end. By applying this important technique, you should find that if your puppy is relaxed and asleep, this calm state will continue.

After a massage, if your puppy is relaxed and sleeping, try not to disturb it, this type of sleep after a gentle massage can be really restorative and very important for a growing puppy and adolescent.

Do not feed your puppy for at least two hours but **do** allow them to drink.

Massage can be highly beneficial and enjoyable for both you and your puppy, be sure you are happy with how to apply the techniques, and a very good adage with massage is 'less is more'.

> Note: The UK Veterinary Act of 1966 prevents any manipulative treatment to be carried out on a dog other than your own except with veterinary consent or referral, therefore, it is illegal for anyone to massage another person's puppy/dog without veterinary consent or referral.

Massage has a profound effect on a puppy's and a dog's whole system. So, if they are unwell, they must not have any form of massage.

When not to massage your puppy:

- Two hours either side of eating.
- Unusually quiet or withdrawn – indicating that they could be ill.
- Has recently vomited or had diarrhoea.
- Unusually lethargic or subdued.
- Off their food.
- You are worried about their health in any way.
- Your puppy does not want to be massaged – never restrain your puppy.

ADDITIONAL HEALTH INDICATORS

Feeling for heat or change over your puppy's body is very important, as it can be a useful indicator of muscle or joint issues, either through injury, a health issue, or even from an overuse or overexercise scenario. Developing this skill will be a great aid for you and your puppy not just for the present but for the rest of their life.

The reason it is important to recognise heat in your puppy and your older dog is that it can be a useful indication of what is happening to your puppy's soft tissue, i.e., muscles, tendons and ligaments, and any joint issues. Heat is produced as a result of an inflammatory response.

Likewise feeling any irregularity within the skeletal system could be important too.

If heat is felt that is different in comparison to the opposing side or greater than felt before, it could demonstrate a stress within an area or an injury, both of which should be considered when looking at what your puppy has been doing and any exercise regimes or playing habits (Figure 9.14 and Table 9.1).

Feeling for heat or change over your puppy's body is very important, as it can be a useful indicator of muscle or joint issues. This could be from injury, a health issue, or even from overexercise. Developing this skill will be a great aid for you and your puppy not just for the present but for the rest of their life.

Place your hands on both sides of your puppy at the same time so they mirror each other. Either the palm or the outer surface of your hand can be used.

Common places where you could feel heat

Figure 9.14 Suggested methodical route for checking for inflammatory heat.

Table 9.1 What to Do If You Find Heat on Your Puppy		
Heat detection	**Possible cause**	**What to do**
Constant heat in a particular area or areas	A chronic or continuing problem causing ongoing inflammation through an ongoing problem, or overuse/overexercise.	Consider if your puppy is doing too much ultra-high impact exercise or even high impact exercise – if so, reduce it immediately. Go to your veterinary surgeon for advice. Follow-up with professional myotherapy and possibly specific corrective exercises.
Heat after exercise in a specific area	An acute and specific problem that may be caused by a recent accident including a collision from play.	Consider if your puppy is doing too much ultra-high or even high impact exercise – if so, reduce it immediately. Go to your veterinary surgeon for advice then if appropriate follow-up with professional myotherapy.

Leave them on the area for ten seconds then move slowly over your puppy's body in a systematic way. You can start this exercise when your puppy first arrives and continue this throughout their life. The more you do it, the more expert you will become, leading to finding and detecting issues early.

Suggested route:

1. Start with your hands flat on your dogs' neck behind the ears.
2. Move down over the front of his shoulder.
3. Put your hands on his shoulders.
4. Move down to over his elbows.
5. Move down his legs.
6. Move back to just behind his shoulders.
7. Move systematically down his back.
8. Hold your hands over his hips/pelvis.
9. Put your hands over the outer surface of his hind legs.
10. Move your hands down to his 'knees' stifles and down his legs.
11. Put your hands inside and behind his hind legs.

10 *Additional useful information*

HOW DO YOU KNOW IF YOUR PUPPY IS BALANCED?

Obviously, puppies are growing and developing into their bodies, so these are guidelines only.

Within the book, we have been looking at how we can encourage our puppies' bodies to become stronger, flexible, and healthy using natural exercise patterns. How will we know if our work has been successful?

When a puppy is growing, it is possible to see if its frame is becoming more secure. Obviously, puppies start with different levels of potential, as some will be born with deformities both skeletal and physiological that will compromise them from becoming as strong as fully healthy puppies. However, this programme *will* help them become the best version of themselves!

Some of the first indicators that your puppy is becoming 'joined up' is that they will demonstrate the following points more than puppies that haven't gone through (and continue into adolescence and beyond) the same programme:

- More balance.
- Better spatial awareness.
- Negotiate new obstacles more easily.

Why is this important? The more balanced and spatially aware they are, the less opportunity for accidents.

PHYSICAL INDICATORS THAT YOUR DEVELOPING PUPPY IS JOINED UP AND BALANCED

1. **The position of the tail in the 'thinking sit'**. When a puppy does a 'thinking sit' (see page 199), its tail should be positioned directly behind them (Figure 10.1).

   When puppies are growing and developing, their sit can look very uneven, often referred to as the 'puppy sit'. This is not the type of sit that is being referred to here. We're looking for the more upright sit that a puppy can use when they are thinking about or studying something. The 'puppy sit' can continue to be seen when they are relaxing or relaxed, but they should ultimately have the ease and ability to **naturally** sit upright, with their paws aligned and their tail lying in line with the rest of their vertebrae.

2. **Wagging tail and back end not moving**. When they are standing and wagging their tail, their hindquarters should remain almost still (Figure 10.2).

DOI: 10.1201/9781003268789-10

Figure 10.1 A puppy in a thinking style 'sit', with its tail positioned in line with the rest of its vertebrae.

Figure 10.2 Puppies wagging their tails with secure hindquarters.

3. **Four square legs**. When they are standing totally naturally, without being positioned, their legs should be opposite each other (lateral, cranial and caudal views, side, back, and front views) (Figure 10.3).

4. **Loose skin**. When you gently lift their skin, it should be easy, soft, and not feel like there is tension or a feeling of it pulling in another direction (Figure 10.4).

5. **Cheeky and a little naughty, as a puppy should be, wanting to play**. Puppies, adolescents, and mature dogs are naturally playful. If this is not the

Figure 10.3 When a puppy is standing naturally, they should stand with their legs squarely opposite each other.

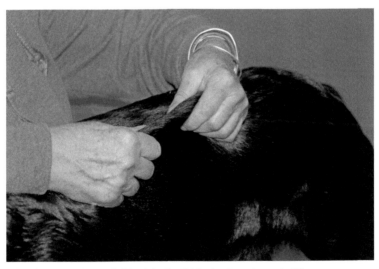

Figure 10.4 When the skin is gently lifted, it should feel soft and easy to lift.

case, then consider that there may be another reason than because they are a 'well-behaved puppy' (see Galen Myotherapy, page 241; Figure 10.5).

6. **Shaking on one spot**. This is something that is a great indicator when your puppy is a little closer to adolescence than as a small puppy. When they shake, their feet should lift up and down, but stay almost in the same position on the ground from the start of the shake to the end of the shake. The shake should also be uninterrupted flowing movement through the body, from nose to tail, one continuous action (Figure 10.6).

Figure 10.5 A puppy should be cheeky and looking to play (and push boundaries!).

Figure 10.6 A dog should be able to shake, lifting and placing its legs back in almost the same position from the beginning to the end of the shake.

This is not a finite list, but a few indicators that could help you understand your puppy's physical stability.

POSTURE

Posture is something that Galen Myotherapy is leading the way in. Specific observations correlate to help the understand the changes in a dog's posture and the physical implications that are leading to overloading, compensation, and other physical changes that can lead to enhanced discomfort and pain, either from a direct or indirect injury or physical adaptation.

It is highly recommended to take a side-on photograph of your puppy/adolescent, or rehomed dog, and place it somewhere visually obvious (a fridge is always a good starting point!). By having an early photo of your dog, it is easier to assess if your dog's posture is altering. Posture can take months to change, but the sooner it can be addressed, the better the recovery and the less danger of additional wear and tear on the whole body.

See page 9 for Maggie and Jess.

See page 190 for muscles and joint damage.

ADDITIONAL INFORMATION TO HELP YOUR PUPPY

GALEN MYOTHERAPY®

Galen Myotherapy®, established in 2002, is a massage and exercise rehabilitation therapy.

It specialises in muscular health, using natural exercise and activities, for dogs of all ages from puppies to veterans.

Its unique methodology for assessing a dog's physical health using physical posture and extensive somatic observations of the dog's appearance, along with correlating behaviours with specific muscular pain, allows targeted treatments using specialised massage techniques, biomechanical assessment, and rehabilitative functional exercise to **help restore the body to a good condition**.

Galen Myotherapy includes puppy assessments, treatments, and physical development with the use of natural exercise and activity.

Galen Myotherapy was the first in our field to develop a choice-led treatment for dogs. Recognising dogs as sentient beings, who lead rich emotional lives and can experience pain. This in-depth understanding of canine behaviour led to the development of their science-based Positive PACT® treatment protocol. As such, Galen Myotherapists never use forced restraint and treatment will always take place at a low level where a dog is most comfortable.

PAIN AND BEHAVIOUR

Puppies or mature dogs can often change their behaviour. This can range from becoming more anxious, having issues with being left alone, and being destructive, aggressive, or even licking, chewing, or biting parts of their body.

Dogs do not display 'pain' or discomfort in the same way as we do. They rarely vocalise, or show normal 'human' indications, but they very often change their behavioural patterns. If your puppy or mature dog has changed their behaviours, please have a veterinary pain assessment.

The problem could be physical but not directly skeletal, and muscle or fascial dysfunction can be difficult to diagnose or interpret. Galen Myotherapy® is one such organisation that specialises in assessing and treating muscular pain or discomfort.

Galen Myotherapy and Galen Myotherapists work with the consent of the dogs' veterinary surgeon. It is always advised to gain a veterinary opinion of your dog's health before seeking any form of complementary therapy.

PUPPY INJURIES

Every action has an equal and opposite reaction (see repetitive strain injuries, page 190), exactly the same as any form of injury. If your puppy has an injury, you must always seek veterinary advice.

If they have been signed off by their vet, even if they cannot find any apparent injury themselves or if other forms of diagnostics have been employed, it is always worth having a soft tissue practitioner expert assess your puppy or dog, as muscle and fascial direct or indirect injury aren't always apparent.

REHOMED PUPPIES OR DOGS

Often, rehomed dogs of all ages have not received the foundation work that is so critical for their physical health and further development. Often, it is viewed that a rehomed dog should have copious amounts of exercise to compensate for what they may have missed in their previous life.

It is strongly advised not to give your rehomed dog excessive exercise but treat them exactly as the Galen Puppy Programme advises. This will help to build and condition them so they can then sustain the rigours of more exercise. Without this type of conditioning, they can suffer very quickly and easily from overloading and repetitive strains.

NUTRITION

It is absolutely vital to feed your puppy the correct food to keep them healthy and strong. It is strongly advised to gain independent, professional advice to ensure a sustainable diet for your puppy. See Gwen Bailey (2021).

COMPETITIVE ACTIVITIES

If you are looking to participate in a future sporting activity with your puppy, it is vitally important not to start actual training too soon. Often, a sport states that a dog can join at a specific age, perhaps 12–18 months. However, starting so young means that a dog has had to start training even before this age, and 18 months is not a guaranteed age for physical (or mental) maturity.

The difference in physical development between a year of age and two is profound. The body has had a chance to develop muscular balance and therefore is potentially more robust with good foundational strength.

Frequently, the very nature of 'sport' means some form of speed and also repetitive actions. Your dog needs established muscle patterning to be engaging and to maintain the correct kinetic chain during physical activity.

For individuals who are looking to compete, it is even more important that you put in the foundation work. If you plan to put your dog's body under additional loading and stress that sport inevitably brings, then developing your dog's skeletal stability through ensuring that the foundation muscles are maintaining their integrity is vital to retain your dog's health and well-being, and to reduce injury.

It should be considered if your dog's build is actually appropriate for physical sport – not every human can run a marathon, be a sprinter, high jumper, or even a gymnast! Dogs are the same.

Maybe consider the less impactful activities that also run competitions:

Scentwork UK is a UK specific branch of an international network of scent or nose work classes for all dogs. This can be a fantastic way of enjoying non-impactful but body and mind developing classes that are progressional and are incredibly bonding for you and your puppy without putting excessive loading through their body, and with good instruction to help the body and mind develop naturally.

Hoopers is a dog sport that's ideal for dogs and owners of all ages and fitness levels. Dogs navigate a course of hoops, barrels, and tunnels with the same pace and excitement as agility. But the courses are flowing and don't involve tight turns.

ORGANISATIONS

Galen Myotherapy (www.galenmyotherapy.co.uk)
Puppy School (www.puppyschool.co.uk/)
PDTE Pet Dog Trainers of Europe (www.pdte.eu/)
An organisation of pet dog trainers, who have been trained by Turid Rugaas (http://en.turid-rugaas.no/)
Scentwork UK (https://scentworkuk.com/)
Canine Hoopers (www.caninehoopersuk.co.uk/)
This is not a finite list, but an indication of the type of organisations that will work with you and your puppy, using more natural and gentle techniques.
For further reading, see the recommended reading, bibliography, and resources.

REFERENCES

Anatomy of the Dog – An Illustrated Text, 4th Edition – ISBN 3 87706 619 4
Guide to the Dissection of the Dog – ISBN 978 1 437 70246 0
The Natural Dog – ISBN 978 0 600 63603 8

Recommended reading

Bailey, G. *The Natural Dog.*

Bradshaw, J. (2011) *Dog Sense*. New York: Basic Books.

Budzinski, C. & Budzinski, A. DogFieldStudy.
http://www.dogfieldstudy.com/en/pulse-study/at-the-heart-of-the-walk

Clark, L. (2019a) *Here to Stay: Training Your Puppy*. Woking: Nielsen.

Clark, L. (2019b) *Here to Stay: Training Your Rescue Dog*. Woking: Nielsen.

Clothier, S. (2002) *Bones Would Rain from the Sky: Deepening Our Relationships with Dogs*. New York: Grand Central Publishing.

Coppinger, R. & Coppinger, L. (2017) *What Is a Dog?* Chicago: University of Chicago Press.

De Wall, F. (2016) *Are We Smart Enough to Know How Smart Animals Are?* Croydon: Granta Books.

Horowitz, A. (2019) *Our Dogs, Ourselves: The Story of a Unique Bond*. New York: Simon & Schuster UK.

Jones, E., MS CDBC, IAABC-ADT, CPDT-KA, CANZ. (2021) What can 'streeties' teach us about companion dogs? *The IAABC Journal* 19(Dog):1–2.

McConnell, P. (2007) *For the Love of a Dog. Understanding Emotion in You and Your Best Friend*. New York: Ballentine Books.

Panagal, S. (2021) *Dog Knows: Learning How to Learn from Dogs*. New Delhi: HarperCollins.

Pilley, J.W. (2013) *Chaser: Unlocking the Genius of the Dog Who Knows a Thousand Words*. New York: Houghton Mifflin Harcourt.

Pryor, K. (2002) *Don't Shoot the Dog: The New Art of Teaching and Training*. Dorking: Surrey.

Robertson, J. & Mead, A. (2013) *Physical Therapy and Massage for the Dog*. Oxford: CRC Press. Taylor & Francis.

Rugass, T. (2005) *My Dog Pulls: What Do I Do?* Wenatchee: Dogwise.

Rugass, T. (2006) *On Talking Terms with Dogs*. Wenatchee: Dogwise.

Zulch, H. & Mills, D. (2012) *Life Skills for Puppies. Laying the Foundation for a Loving, Lasting Relationship*. Dorchester: Hubble & Hattie.

Zulch, H. & Mills, D. (2015) *Helping Minds Meet: Skills for a Better Life with Your Dog*. Dorchester: Hubble & Hattie.

Bibliography

Anderson, K.L., O'Neill, D.G., Brodbelt, D.C., Church, D.B., Meeson, R.L., Sargan, D., Summers, J.F., Zulch, H. & Collins, L.M. (2018) Prevalence, duration and risk factors for appendicular osteoarthritis in a UK dog population under primary veterinary care. *Scientific Reports* 8:5641.

Bailey, G. (2021) *The Natural Dog: The New Approach to Achieving a Happy, Healthy Hound*. London: Hamlyn.

Batson, A. (2021) Cavorting canines. Seminar presented Active Chiens, October 2021.

Bordoni, B., Morabito, B., Mitrano, R., Simonelli, M. & Toccafondi, A. (2018) The anatomical relationships of the tongue with the body system. *Cureus* 10:e3695.

Bradshaw, J. (2011) *In Defence of Dogs*. London: Penguin.

Craig, A.D. (2003) A new view of pain as a homeostatic emotion. *Trends in Neurosciences* 26(3):303–307.

Dietz, L., Arnold, A.K., Coerlich-Jansson, V.C. & Vinke, C.M. (2018) The importance of early life experiences for the development of behavioural disorders in domestic dogs. *Behaviour* 155:83–114.

Harari, Y.N. (2011) *Sapiens: A Brief History of Humankind*. London: Vintage.

Korda, P. (1972) Epimeletic vomiting in female dogs during the rearing process of their puppies. *Acta Neurobiol Exp (Wars)* 32(3):733–747.

Majunder, S.S., Chatterjee, A. & Bhadra, A. (2014) A dog's day with humans – time activity budget of free ranging dogs in India. *Current Science* 106(6):874–878.

McConnell, P.B. (2007) *For the Love of a Dog: Understanding Emotion in You and Your Best Friend*. New York: Ballantine Books.

McMillan, F.D. (2021) *Mental Health and Well-Being in Animals*. 2nd edition. Oxford: CABI.

Mills, D., Braem Dube, M. & Zulch, H. (2013) *Stress and Pheromanotherapy in Small Animal Clinical Behaviour*. Oxford: Wiley-Blackwell.

Mills, D., Demontigny-Bédard, I., Gruen, M., Klinck, M.P., McPeake, K.J., Barcelos, A.M., Hewison, L., Haevermaet, H.V., Denenberg, S., Hauser, H., Koch, C., Ballantyne, K., Wilson, C., Chirantana V., Mathkari, C.V., Pounder, P., Garcia, E., Darder, P., Fatjó, J. & Levine, E. (2020) Pain and problem behaviour in cats and dogs. *Animals* 10:318.

Morton, C.M., Reid, J., Scott, J., Holton, L.L. & Nolan, A.M. (2005) Application of a scaling model to establish and validate level pain scale for assessment of acute pain in dogs. *American Journal of Veterinary Research* 66(12):2154–2166.

Overall, K.L. (2001) Where does our behaviour come from? *Journal of Bioscience* 26(5):561–570.

Overall, K.L. (2013) *Manual of Clinical Behavioural Medicine for Cats and Dogs*. London: Elsevier.

Packer, R.M.A., O'Neill, D.G., Fletcher, F., Farnworth, M.J. (2019) Great expectations, inconvenient truths, and the paradoxes of the dog-owner relationship for owners of brachycephalic dogs. *PLoS ONE* 14(7):e0219918. https://doi.org/10.1371/journal.pone.0219918

Salt, C., Morris, P.J., Wilson, D., Lund, E.M. & Alexander, J. (2018) Association between life span and body condition in neutered client-owned dogs. *Journal of Veterinary Internal Medicine* 1(2), https://onlinelibrary.wiley.com/doi/epdf/10.1111/jvim.15367

Schneider, S.M. (2012) *The Science of Consequences: How They Affect Genes, Change the Brain and Impact Our World*. New York: Prometheus Books.

Scott, J.P. & Fuller, J.L. (1965) *Genetics and the Social Behaviour of the Dog*. Chicago: University of Chicago Press.

Seligman, E.P. (1972) *Learned Helplessness*. Philadelphia, PA: Departments of Psychiatry and Psychology, University of Pennsylvania.

Warren, C. (2016) *What the Dog Knows: Scent, Science, and the Amazing Ways Dogs Perceive the World*. Australia and New Zealand: Scribe.

Resources

ABTC Animal Behaviour Training Council (https://abtc.org.uk)

The Animal Behaviour and Training Council is the regulatory body that represents animal trainers, training instructors, and animal behaviour therapists to both public and legislative bodies. It sets and maintains the standards of knowledge and practical skills needed to be an animal trainer, training instructor, or animal behaviour therapist, and it maintains the national registers of appropriately qualified animal trainers and animal behaviourists.

APBC Association of Pet Behaviour Counsellors (www.apbc.org.uk)

The APBC is an international network of experienced and qualified animal behaviour counsellors who work on referrals from veterinary surgeons to treat behaviour problems in dogs, cats, birds, rabbits, horses, and other animals.

APDT Association of Pet Dog Trainers (https://apdt.co.uk)

The APDT is a not-for-profit organisation to promote positive training skills that improve the welfare of dogs and promotes the competence of dog owners.

Canine Hoopers (www.caninehoopersuk.co.uk/)

The home of hoopers, training classes, competitions, awards, and instructor courses in the UK.

Galen Canine Myotherapy (www.galenmyotherapy.co.uk)

Galen (Canine) Myotherapy is a specialist branch of massage therapy which promotes health and treats chronic muscular pain, and supports muscle health in dogs of all ages, through unique massage techniques, and rehabilitative exercise management. Galen Myotherapists use the scientifically supported choice-led treatment protocol called Positive P.A.C.T.®. Their clinical assessments include the Galen Comfort Scale™, which is used to correlate a dog's pain and behaviour, and the dog's posture and postural changes.

FCAB Fellowship of Clinical Animal Behaviour Technicians (https://fabclinicians.org)

The Fellowship of Animal Behaviour Clinicians is a professional body to forward the interests of Certificated Clinical Animal Behaviourists (CCABs) and those training to become Certificated Clinical Animal Behaviourists.

KPA Karen Pryor Academy (https://karenpyroracademy.com)

Founded by Karen Pryor, the KPA academy is an international organisation offering online and practical animal training and instructor courses. Specialises in clicker training.

PACT Professional Association of Canine Trainers (www.pact-dogs.com)

PACT dogs is a nationally recognised and externally assessed UK dog trainer training and accreditation and membership body.

PDTE Pet Dog Trainers of Europe (www.pdte.eu/)

An organisation of pet dog trainers, who have been trained by Turid Rugaas (http://en.turid-rugaas.no/).

Puppy School (www.puppyschool.co.uk/)

Scentwork UK (https://scentworkuk.com/)

Promoting scent work as a dog sport and training activity.

TCTBS The Canine Training and Behaviour Society (www.tctbs.co.uk)

The Canine Behaviour and Training Society is a non-profit, professional body of canine behaviourists and dog training instructors.

This is not a finite list, but an indication of the type of organisations that will work with you and your puppy, using more natural and gentle techniques.

Puppy Parties – be aware before taking your puppy to a Puppy Party, that the trainer or organiser is correctly qualified, the numbers of puppies attending are limited to very few, whatever size room it is being held, and that the flooring is non-slip.

Other methods of socialising your puppy can be more successful and much less stressful, and potentially damaging to your puppy.

Index

Note: Locators in *italics* represent figures and **bold** indicate tables in the text.